Modeling and Analysis of Local Area Networks

Modeling and Analysis of Local Area Networks

Paul J. Fortier
George R. Desrochers

CRC Press
Boca Raton   Ann Arbor   Boston

Library of Congress Catalog Card Number 89-83757

This book represents information obtained from authentic and highly regarded sources. Reprinted material is quoted with permission, and sources are indicated. A wide variety of references are listed. Every reasonable effort has been made to give reliable data and information, but the authors and the publisher cannot assume responsibility for the validity of all materials or for the consequences of their use.

All rights reserved. This book, or any parts thereof, may not be reproduced in any form without written consent from the publisher.

Direct all inquiries to CRC Press, Inc., 2000 Corporate Blvd., N.W., Boca Raton, Florida, 33431.

© 1990 by Multiscience Press, Inc.

International Standard Book Number 0-8493-7405-7
Printed in the United States of America

Contents

	Preface	ix
1	**Introduction**	1
	Models 2	
	Model Construction 3	
	Tools 6	
	Simulation Modeling 8	
	Test Beds 9	
	Operational Analysis 10	
	Networking and LANs 12	
	The Need for Evaluation 13	
	Book Summary 16	
2	**Local Area Networks**	17
	Local Area Networks: Their Uses 19	
	Lan Technology 21	
	Regular Topology 26	
	Hybrids 28	
	Interface Units 28	
	Control Protocols 30	
	Contention Based 31	
	Reservation Based 34	
	Sequential Based 35	
	Systems Management 37	
	Database Manager 39	
	Summary 40	
	References 41	
3	**Modeling of Local Area Networks**	43
	Techniques 47	
	Summary 48	

vi Modeling and Analysis of Local Area Networks

4   Probability and Statistics — 49
Random Variables 55
Jointly Distributed Random Variables 55
Probability Distributions 56
Densities 58
Expectation 61
Some Example Probability Distributions 68
Summary 83
References 83

5   Simulation Analysis — 85
Time Control 88
Systems and Modeling 89
Discrete Models 89
Continuous Modeling 91
Queuing Modeling 93
Combined Modeling 93
Hybrid Modeling 94
Simulation Languages 94
GASP IV 94
GPSS 98
SIMSCRIPT 103
SLAM II 103
Applications of Simulation 109
The Simulation Program 110
Summary 115
References 115

6   Queuing Theory — 117
Stochastic Processes 117
Markov Processes 125
Queuing Systems 127
Networks of Queues 147
Estimating Parameters and Distributions 155
Summary 161
References 161

7   Computational Methods for Queuing Network Solutions — 163
Central Server Model 163
Mean Value Analysis 170
Operational Analysis 173
Summary 179
References 180

Contents   vii

8   Hardware Test Beds   **181**
       Derivation of Performance Evaluation Parameters   187
       Network Performance Tests   190

9   LAN Analysis   **195**
       Introduction   195
       Analytical Modeling Examples   198
       HXDP Model   198
       Graphic Outputs   202
       Token Bus Distributed System   204
       Summary   215
       References   217

10   MALAN   **219**
       Introduction   219
       Simulating Local Area Networks   219
       Computer Networks (the Model)   220
       Protocols   226
       Transmission Error Detection   231
       Events   233
       The Malan Model Structure   236
       Malan Simulator Overview   237
       Next Event Simulation   238
       General Model Implementation   239
       Model Implementation   240
       Malan Interactive Simulation Interface   242
       Malan Model Implementation   243
       Arrival Module   245
       Arbitrator Module   253
       Use Module   255
       The Organization of the Use Module   255
       Analysis Module   261
       MALAN IMPLEMENTATION   267
       REFERENCES   270

       Appendix A   HXDP Calculations   **285**

       Appendix B   Token Bus Computations   **297**

       References   **305**

       Index   **307**

Preface

Local Area Networks (LANs) have been around in various forms since the late 1960s and early 1970s. Since that time, they have found their way into many systems that affect our everyday life. For instance, LANs have been designed into automobiles to link the various sensors and microprocessors now found in cars to monitor everything from air temperature, moisture, fuel-to-air mixtures, fuel economy, and emission control to even determining if the driver is sober enough to drive. The network has become an integral part of our computer and information management technologies.

In order for us as LAN designers, users, purchasers, and researchers to make sound decisions regarding local area networks, we need to have sufficient information on their characteristics and operation.

Books have been published that define what a LAN is, what it can do for us, and how it performs, although texts that address LANs as a component of a system, needing to be analyzed and evaluated are limited. This text aims at filling that void. The goals of this text are to review the state of technology for LANs from a hardware and software perspective, develop a set of metrics that can be used to evaluate LANs for end applications, and investigate methodologies for evaluating LANs from these perspectives.

Included in this book are LAN evaluation techniques utilizing analysis, operational analysis, hardware testbeds, and simulations. Simulations will be stressed in greater detail and a tool available for evaluating LANs performance — which we call MALAN — will be presented and the details of its structure developed.

The simulator discussed can be obtained from the authors for use by researchers, users, designers, or evaluators, to aid them in their LAN modeling endeavors.

Paul Fortier
255 Fairview
Portsmouth, RI 02871

George Desrochers
Georgetown, MA

1

Introduction

Presently there is a great deal of interest and activity in the design and use of distributed local area networks (LANs). These networks are being researched, developed, produced, and marketed by individuals and organizations from government, industry, and academia. These activities are motivated by the rapidly changing technologies of devices and systems, increased performance requirements, increasing complexity and sophistication of interconnections and control, the constant demand for improved reliability and availability, and the increasing reliance of organizations on the use of computer facilities in all aspects of business.

Local area networks provide more features than are presently available in a single, large time-sharing system. Specifically, those features are the sharing of resources on a much more global scale, the fulfillment of system requirements such as expandability, flexibility, availability, reliability, reconfigurability, fault tolerance, graceful degradation, responsiveness, speed, throughput capacity, logical complexity, and ease of development (modularity). Another appealing feature of distributed systems is their ability to bring the computing power to the user without sacrificing the ability to get at all the available information assets from the business.

The optimal design and/or selection of a LAN is, therefore, of the utmost importance if the computing facility is to provide the new and improved services. But how does one go about doing this? What techniques and tools are available for this purpose? These are among questions that this book will address for the LAN designer, purchaser, or student. It is set up to cover the essentials of modeling and analysis of local area network communication environments. Covered are the basic technologies associated with LANs, details of modeling techniques used to study LANs, and the description of a software tool that has been used to model LANs.

The software tool described can be used to model a variety of LAN topologies and control environments as well as the details of the underlying policies and mechanisms that govern LAN operations. It has been developed so that it allows tailoring to specific LAN characteristics and enables growth into the modeling of the upper layers of the LAN environment (i.e., the operating system, information manager, or applications processes).

But before we get into details of local area networks and their performance assessment, we must define the boundaries of what is being studied. What are models and how can they be used to provide analysis of systems?

Models provide a tool for users to define a system and its problems in a concise fashion; they provide vehicles to ascertain critical elements, components, and issues; they provide a means to assess designs or to synthesize and evaluate proposed solutions; and they can be used as predictions to forecast and aid in planning future enhancements or developments. In short, they provide a laboratory environment in which to study a system even before it exists or without actually affecting an actual implementation. In this light models are descriptions of systems. Models typically are developed based on theoretical laws and principles. They may be physical models (scaled replicas), mathematical equations and relations (abstractions), or graphical representations. Models are only as good as the information put into them. That is, modeling of a system is easier and typically better if:

- Physical laws are available that can be used to describe it.
- Pictorial representations can be made to provide better understanding of the model.
- The system's inputs, elements, and outputs are of manageable magnitude.

These all provide a means to realize models, but the problem typically is that we do not have clean and clear physical laws to go on, interactions can be very difficult to describe, randomness of the system or users causes problems, and policies that drive processes are hard to quantify. What typically transpires is that a "faithful" model of a system is constructed, one that provides insight into a critical aspect of a system, not all of its components. That is, we typically model a slice of the real world system. What this implies is that the model is an abstraction of the real world system under study. With all abstractions, one must decide what elements of the real world to include in the abstraction and which ones are important to realize as a "faithful" model. What we are talking about here is intuition; that is, how well can a modeler select the significant elements, how well can they be defined, and how well can the interaction of these significant elements within themselves, among themselves, and with the outside be defined?

MODELS

As stated above, a model is an abstraction of a system. The realism of the model is based on the level of abstraction applied. That is, if we know all there is about a system and are willing to pay for the complexity of building a true model, the abstraction is near nil. On the other hand, in most cases we wish to abstract the view we take of a system to simplify the complexities. We wish to build a model that focuses on some element(s) of interest and leave the rest of the system as only an interface with no detail beyond proper inputs and outputs.

The "system," as we have been calling it, is the real world that we wish to model (for instance, a bank teller machine, a car wash, or some other tangible item or process). Pictorially (see Figure 1-1) a system is considered to be a unified group of objects united to perform some set function or process, whereas a model is an abstraction of the system that extracts out the important items and their interactions.

The basic notion of this is that a model is a modeler's subjective view of the system. This view defines what is important, what the purpose is, detail, boundaries, etc. The modeler must understand the system in order to provide a faithful perspective of its important features and in order to make the model useful.

MODEL CONSTRUCTION

In order to construct a model, we as modelers must follow predictable methodologies in order to derive correct representations. The methodology typically used consists of top-down decomposition and is pertinent to the goal at each step of being able to define the purpose of the model or its component and, based on this purpose, to derive the boundaries of the system or component and develop the level of modeling detail. This iterative method of developing purpose, boundaries, and modeling level smoothes out the rough or undefinable edges of the actual system or component, thereby focusing on the critical elements of it.

The model's inputs are derived from the system under study as well as from the performance measures we wish to extract. That is, the type of inputs are detailed not only from the physical system but through the model's intended use (this provides the experimental nature of the model). For instance, in an automated teller

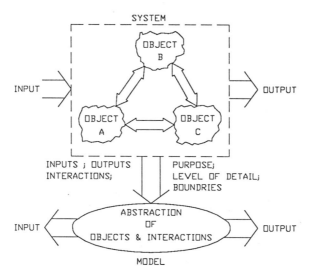

Figure 1-1 Abstractions of objects and interactions.

machine, we wish to study the usefulness of fast service features such as providing cash or set amounts of funds quickly after the amount has been typed in.

The model would be the bank teller machine, its interface, and a model (analytical, simulation) of the internal process. The experiment would be to have users (experimenters) use the model as they would a real system and measure its effectiveness. The measures would deal with the intent of the design. That is, we would monitor which type of cash access feature was used over another, etc.

The definition of the wanted performance measures drive the design and/or redesign of the model. In reality, the entire process of formulating and building a model of a real system occurs interactively. As insight is gained about the real system through studying it for modeling purposes, new design approaches and better models and components are derived. This process of iteration continues until the modeler has achieved a level of detail consistent with the view of the real system intended in the model-purpose development phase. The level of detail indicates the importance of each component in the modeler's eye as points that are to be evaluated.

To reiterate, the methodology for developing and using a model of a system are:

1. Define the problem to be studied as well as the criteria for analysis.
2. Define and/or refine the model of the system (includes development of abstractions of the system into mathematical, logical, and procedural relationships).
3. Collect data for input to the model (define the outside would and what must be fed to or taken from model to "simulate" that world).
4. Select a modeling tool—prepare and augment the model for tool implementation.
5. Verify that the tool implementation is an accurate reflection of the model.
6. Validate that the tool implementation provides the desired accuracy or correspondence with the real world system being modeled.
7. Experiment with the model to obtain performance measures.
8. Analyze the tool results.
9. Use these findings to derive designs and improvements for the real world system.

Although some of these "steps" were defined in the previous paragraphs, they will be readdressed here in the context of the methodology.

The first task in the methodology is to determine what the scope of the problem is and if this real world system is amenable to modeling. This task consists of clearly defining the problem and explicitly delineating the objectives of the investigation. This task may need to be reevaluated during the entire model construction phase because of the nature of modeling. That is, as more insight comes into the process, a better model, albeit a different one, may arise. This involves redefinition of questions being asked and the evolution of a new problem definition.

Once a "problem definition" has been formulated, the task of defining and refining a model of this real world problem space can ensue. The model typically is made up of multiple sections that are both static and dynamic. They define elements of the system (static), their characteristics, and the ways in which these elements interact over time to adjust or reflect the state of the real system over time. As indicated earlier, this process of formulating a model is largely dependent on the modeler's knowledge, understanding, and expertise (art versus science). The modeler extracts the "essence" of the real world system without encasing superfluous detail. This notion is that of capturing the crucial (most important) aspects of the system without undue complexity, though with enough to realistically reflect the germane aspects of the real system. The amount of detail to include in a model is based mainly on its purpose. For example, if we wish to study the user transaction ratio and types on an automated teller machine, we only need model the machine as a consumer of all transaction times and their types. We need not model the machine and its interactions with a parent database in any detail but only from a gross exterior user level.

The process of developing the model from the problem statement is iterative and time-consuming. However, a fallout of this phase is the definition of input data requirements. Added work typically will be required to gather the defined data values to drive the model. Many times in model development data inputs must be hypothesized or based on preliminary analysis, or the data may not require exact values for good modeling. The sensitivity of the model is turned into some executable or analytical form and the data can be analyzed as to its effects.

Once the planning and development of a model and data inputs have been performed, the next task is to turn it into an analytical or executable form. The modeling tool selected drives much of the remainder of the work. Available tools include simulation, analytical modeling, test beds, and operational analysis. Each of these modeling tools has its pros and cons. Simulation allows for a wide range of examinations based on the modeler's expertise, and analytical analysis provides best, worst, and average analysis but only to the extent of the modeler's ability to define the system under study mathematically. Test beds provide a means to test the model on real hardware components of a system, but they are very expensive and cumbersome. The last, operational analysis, requires that we have the real system available and that we can get it to perform the desired study. This is not always an available alternative in complex systems. In any case, the tool selected will determine how the modeler develops the model, its inputs, and its experimental payoff.

Once a model and a modeling tool to implement it have been developed, the modeler develops the executable model. Once developed, this model must be verified to determine if it accurately reflects the intended real world system under study. Verification typically is done by manually checking that the model's computational results match those of the implementation. That is, do the abstract model and implemented model do the same thing and provide consistent results?

Akin to verification is validation. Validation deals with determining if the model's implementation provides an accurate depiction of the real world system being modeled. Testing for accuracy typically consists of a comparison of the model and system structures against each other and a comparison of model tool inputs, outputs, and processes versus the real system for some known boundaries. If they meet some experimental or modeling variance criteria, we deem the model an accurate representation of the system. If not, the deficiencies must be found, corrected, and the model revalidated until concurrence is achieved.

Once the tool implementation of the model has been verified and validated, the modelers can perform the project's intended experiments. This phase is the one in which the model's original limitations can be stretched and new insights into the real system's intricacies can be gained. The limitations on experimentation are directly related to the tool chosen—simulation is most flexible, followed by test beds, analytical analysis, and operational analysis.

Once experimentation is complete, an ongoing analysis of results is actively performed. This phase deals with collecting and analyzing experimentally generated data to gain further insight into the system under study. Based on the results generated, the modeler feeds into the decision-making process for the real world system, potentially changing its structure and operations based on the model's findings. A study is deemed successful when the modeling effort provides some useful data to drive the end product. The outputs can solidify a concept or notion about the system, define a deficiency, provide insight into improvements, or corroborate on other information about the system. Modeling is a useful tool with which to analyze complex environments.

TOOLS

As was briefly indicated in the previous section, there are major classes of modeling tools in use today: analytical, simulation, operational analysis, and test beds. Each has its niche in the modeler's repertoire of tools and is used for varying reasons, as will be indicated later.

Analytical

Analytical tools have been used as an implementation technique for models for quite some time, the main reason being that they work. Analytical implementations of models rely on the ability of the modeler to describe a model in mathematical terms. Typically, if a system can be viewed as a collection of queues with service, wait, and analytical times defined analytically, queuing analysis can be applied to solve the problem.

Some of the reasons why analytical models are chosen as a modeling tool are as follows:

Introduction 7

1. Analytical models capture more salient features of systems; that is, most systems can be represented as queuing delays, service times, arrival times, etc., and, therefore, we can model from this perspective, leaving out details.
2. Assumptions or analysis are realistic.
3. Algorithms to solve queuing equations are available in machine form to speed analysis.

What Graham really is saying is that queuing models provide an easy and concise means to develop analysis of queuing-based systems. Queues are waiting lines and queuing theory is the study of waiting line dynamics.

In queuing analysis at the simplest level (one queue figure), there is a queue (waiting line) that is being fed by incoming customers (arrival rate); the queue is operated on by a server who extracts customers out of the queue according to some service rate (see Figure 1-2).

The queue operates as follows: An arrival comes into the queue, and if the server is busy, the customer is put in a waiting facility (the queue) unless the queue is full, in which case the customer is rejected (no room to wait). On the other hand, if the queue is empty, the customer is brought into the service location and is delayed the service rate time. The customer then departs the queue.

In order to analyze this phenomenon we need to have notation and analytical means (theories) with which to manipulate the notation. Additionally, to determine the usefulness of the technique, we need to know what can be analyzed and what type of measure is derived from the queue.

The notation used (see Figure 1-2) to describe the queue phenomenon is as follows: The arrival distribution defines the arrival patterns of customers into the queue. These are defined by a random variable that defines the interarrival time. A typically used measure is the Poisson arrival process defined as : P[arrival $\leq$ time]=$1-e^{-\lambda t}$; where the average arrival rate is λ. The queue is defined as a storage reservoir for customers. Additionally, the policy it uses for accepting and removing customers is also defined. Examples of queuing disciplines typically used are first in first out (FIFO) and last in first out (LIFO). The last main component of the queue description is the service policy, which is the method by which customers are accepted for service and the length of the service. This service time is described by a distribution, a random variable. A typical service time distribution is the random service given by $Ws(t) = 1-e^{-\mu t}$, where $t > 0$, and the symbol μ is reserved

Figure 1-2 Simple model of a queing system.

to describe this common distribution for its average service rate. The distributions used to describe the arrival rate and service ratios are many and variable; for example, the exponential, general, Erlang, deterministic, or hyper-exponential can be used. The Kendall notation was developed to describe what type of queue is being examined. The form of this notation is:

A/B/c/K/m/Z

where A specifies the interarrival time distribution, B the service time distribution, c the number of servers, K the system capacity, m the number in the source, and Z the queue discipline.

This type of analysis can be used to generate statistics on average wait time, average length of the queue, average service time, traffic intensity, server utilization, mean time in system, and various probability of wait times and expected service and wait times. More details on this modeling and analysis technique will be presented in Chapter 4.

SIMULATION MODELING

Simulation as a modeler's tool has been used for a long time and has been applied to the modeling and analysis of many systems, for example, business, economics, marketing, education, politics, social sciences, behavioral sciences, international relations, transportation, law enforcement, urban studies, global systems, computers, factories, and too many more to mention. Simulation lends itself to such a variety of problems because of its flexibility. It is a dynamic tool that provides the modeler with the ability to define models of systems and put them into action. It provides a laboratory in which to study a myriad of issues associated with a system without disturbing the actual system. A wide range of experiments can be performed in a very controlled environment; time can be compressed, allowing the study of otherwise unobservable phenomena, and sensitivity analysis can be done on all components.

However, simulation can have its drawbacks. Model development can become expensive and require extensive time to perform, assumptions made may become critical and cause a bias on the model or even make it leave the bounds of reality, and finally, the model may become too cumbersome to use and initialize effectively if it is allowed to grow unconstrained. To prevent many of these ill affects, the modeler must follow strict policies of formulation, construction, and use. These will minimize the bad effects while maximizing the benefits of simulation.

There are many simulations based on the system being studied. Basically there are four classes of simulation models: continuous, discrete, queuing, hybrid. These four techniques provide the necessary methods to model most systems of interest. A continuous model is one whose processing state changes in time based on time varying signals or variables. Discrete simulation relies instead on event conditions to change state. Queuing-based simulations provide dynamic means to construct and analyze queuing-based systems. They dynamically model the mathematical occurrences analyzed in analytical techniques.

Simulations are performed for systems based on the system's fit to simulation. That is, before we simulate a real world system model, we must determine that the problem requires or is amenable to simulation. The important factors to consider are the cost, the feasibility of conducting useful experimentations, and the possibility of mathematical or other forms of analysis. Once simulation is deemed a viable candidate for model implementation, a formal model "tuned" to the form of available simulation tools must be performed. Upon completion of a model specification, the computer program that converts this model into executable form must be developed. Finally, once the computer model is verified and validated, the modeler can experiment with the simulation to aid in the study of the real world system.

Many languages are available to the modeler for use in developing the computer executable version of a model: GPSS, Q-gert, Simscript, Slam, and Network 2.5. The choice of language is based on the users' needs and preferences, since any of these will provide a usable modeling tool for implementing a simulation. Details of these and the advantages of other aspects of simulation are addressed in Chapters 3 and 5.

TEST BEDS

Test beds, as indicated previously, are abstractions of systems and are used to study system components and interactions to gain further insight into the essence of the real system. They are built of prototypes and pieces of real system components and are used to provide insight into the workings of an element(s) of a system. The important feature of a test bed that it only focuses on a subset of the total system. That is, the important aspect that we wish to study, refine, or develop is the aspect implemented in the test bed. All other aspects have stubs that provide their stimulus or extract their load but are not themselves complete components, just simulated pieces. The test bed provides a realistic hardware-software environment with which to test out components without having the ultimate system. The test bed provides a means to improve the understanding of the functional requirements and operational behavior of the system. It supplies measurements from which quantitative results about the system can be derived. It provides an integrated environment in which the interrelationships of solutions to system problems can be evaluated. Finally it provides an environment in which design decisions can be based on both theoretical and empirical studies.

What all this indicates, again, is that, as with simulation and analytical tools, the test bed provides a laboratory environment in which the modeled system components can be experimented with, studied, and evaluated from many angles. However, test beds have their limitations in that they cost more and are limited to only modeling systems amenable to such environments. For example, we probably would not model a complex system in a test bed. We would instead consider analytical simulation as a first pass and use a test bed between the initial concept and final design. This will be more evident as we continue in our discussion here and in Chapter 8 where test beds are discussed in much greater detail.

A test bed is made up of three components: an experimental subsystem, a monitoring subsystem, and a simulation-stimulation subsystem. The experimental subsystem is the collection of real system components and/or prototypes that we wish to model and experiment with. The monitoring subsystem consists of interfaces to the experimental system to extract raw data and a support component to collate and analyze the collected information. The simulation-stimulation subsystem provides the hooks and handles necessary to provide the experimenter with real world system inputs and outputs to provide a realistic experimentation environment.

With these elements a test bed can provide a flexible and modular vehicle with which to experiment with a wide range of different system stimulus, configurations, and applications. The test bed approach provides a method to investigate system aspects that is complementary to simulation and analytical methods.

Decisions about using a test bed over the latter methods are driven mainly by the cost associated with development and the actual benefits that can be realized by such implementations. Additionally, the test bed results are only as good as the monitor's ability to extract and analyze the real world phenomena occurring and the simulation-stimulation component's ability to reflect a realistic interface with the environment.

Test beds in the context of local area networks can and have been used to analyze a wide range of components. The limitation to flexibility in analyzing very diverse structures and implementations has and will continue to be in the cost associated with constructing a test bed. In the context of a LAN, the test bed must implement a large portion of the network hardware and software to be useful. By doing this, however, the modeler is limited to studying this single configuration, not many. It will be seen in later sections what these limitations and benefits are.

OPERATIONAL ANALYSIS

The last tool from a modeler's perspective is operational analysis. In this technique, the modeler is not concerned as much with an abstraction of the system, but with how to extract from the real system information upon which to develop the same analysis of potential solutions that is provided with the other models.

Operational analysis is concerned with extracting information from a working system that is used to develop projections about the system's future operations.

Additionally, this method can be used by the other three techniques to derive meaningful information that can be fed into their processes or used to verify or validate their operations.

Operational analysis deals with the measurement and evaluation of an actual system in operation. Measurement is concerned with instrumenting the system to extract the information. The means to perform this uses hardware and/or software monitors.

Hardware monitors consist of a set of probes or sensors, a logic-sensing device, a set of counters, and a display or recording unit. The probes monitor the state of the chosen system points. Typically, probes can be "programmed" to trigger on a specific event, thereby providing the ability to trace specific occurrences within a system.

The logic-sensing subsystem is used to interpret the raw input data being probed into meaningful information items. The counters are used to set sampling rates on other activities requiring timed intervals. The last component records and displays the information as it is sensed and reduced. Further assistance could be added to analyze the information further. The ability to perform effective operational analysis is directly dependent on the hardware and software monitors' ability to extract information. The hardware monitor is only as effective as its ability to be hooked into the system without causing undue disturbance. The problem is that the hardware-based monitor cannot, in a computer system, sense software-related events effectively. The interaction of software and system hardware together will provide much more effective data for operational analysis to be performed on. Software monitors typically provide event tracing or sampling styles. Event trace monitors are composed of a set of system routines that are evoked on specific software occurrences such as CPU interrupts, scheduling phases, dispatching, lockouts, I/O access, etc. The software monitor is triggered on these events and records pertinent information on system status. The information can include the event triggered at the time, what process had control of the CPU prior to the event, and the state of the CPU (registers, conditions, etc.). This data can reveal much insight as to which programs have the most access to the CPU, how much time is spent in system service overhead, device queue lengths, and many other significant events.

The combination of the hardware and software monitors provides the analyst with a rich set of data on which to perform analysis. Typical computations deal with computing various means and variances of usages of devices and software and plotting relative frequencies of access and use.

The measurements and computations performed at this level only model present system performance. The operational analyst must use these measures to extend performance and to postulate new boundaries based on extending the data into unknown regions and performing computations based on the projected data. Using these techniques, the analyst can suggest changes and improvements and predict their impact based on real information.

NETWORKING AND LANS

The term "network" can mean many different things. It can imply an interconnection of railway tracks for the rail network, highways and streets for transportation networks, telephone lines and switching centers for the phone network, or the interconnection of service centers, businesses, etc., to form a network. It refers to the means to tie together various resources so that they may operate as a group realizing the benefits of numbers and communications in such a group. In computer terms a network is a combination of interconnected equipment and programs used for moving information between points (nodes) in the network where it may be generated, processed, stored, or used in whatever fashion is deemed appropriate. The interconnection may take on many forms such as dedicated links, shared links, telephone lines, microwave links, and satellite links. Networks in this sense form a loose coalition of devices that share information. This was one of the first uses of a network, although it was not the last. Users found that the network could offer more than just information sharing; they could offer other services for remote job execution and ultimately distributed computing.

The early concept of a network was of a loose binding together of devices or resources for sharing. An early computer communications network that exhibited these traits was Arpanet. Arpanet was first brought online in 1969 as a research tool to investigate long-haul network issues and research and development solutions. It presently has over 100 locations strewn over the country connecting thousands of computers over local area networks, metropolitan area networks, and other wide area networks. Arpanet provided the vehicle for early research into communications protocols, dealing with congestion, control, routing, addressing, remote invocation, distributed computing, distributed operating systems and services, and many other areas.

The reasons for using networks such as Arpanet were to provide greater availability and access to a wider range of devices. Early applications of computers dealt with performing engineering tasks and major data processing functions. As the technology of computers changed, and as researchers and users alike added more and more applications, information access and manipulation took on greater emphasis.

The networks of the early times provided the necessary information exchange services but were limited to basically just this service. The information availability stimulated more imaginative uses of this information. As this occurred and the technology of networks improved, new applications arose. These new applications not only used information exchange but also remote job execution. It began as simply sending a batch job down the link to a less busy host, having the job completed there, and then shipping the results back to the originator.

This sufficed for awhile, but it still did not provide the real time or interactive environments that users were beginning to become accustomed to, including more advanced protocols and network operating systems to provide further services for remote job invocation and synchronization. The era of the local area network was coming. The wide area networks' biggest shortfall was in throughput or turnaround time for jobs and interprocessor communications. Because of the wide distances, delays of seconds were commonplace and caused much added overhead in performing otherwise simple tasks. Network designers saw the need to provide another link in the network, the local area network.

Local area networks began to show up in the early to mid-1970s as research activities in universities. It was not until Ethernet was released in the mid-1970s as a product did LANs become more widely available. Since that time, numerous LAN designs have been produced to fit an extremely wide spectrum of user requirements. Additionally, standards have evolved, providing basic LANs and their services to a greater number of users.

Local area networks are finding their way into all aspects of modern society. We find them in our homes, automobiles, banking, schools, businesses, government, and industry. There are not too many aspects of information exchange and data processing in which a LAN cannot be found. Local area networks and their associated technologies represent one of great growth areas of the 1980s and 1990s. As more and more LANs become available, so will new products and uses for them. LANs are used to connect all the personal computers in offices. They are used in this environment to send memoranda, issue directives, schedule meetings, transmit documents, and process large volumes of data concurrently at many sites.

LANs are used to link factory robots together with area and factory controllers. They provide sensor data, control data, and feedback to the control centers, while at the same time providing a vehicle to issue production changes and notices to users and robots alike. A fine example of a local area network providing diverse services to the users is seen in Walt Disney World. Disney uses LANs and computers to monitor all aspects of services, including fire protection, scheduling, ride management, online information, security, personnel services, and a plethora of other park management functions. Large banks, such as the World Bank, have adopted LANs as the means to interconnect their various local sites into smaller networks linked together by wide area networks. However, the LAN is not for everyone.

THE NEED FOR EVALUATION

Selecting a local area network architecture and system that will provide the optimum service to the users requires up front analysis and knowledge. As indicated, a LAN is a productivity-enhancing tool, but like other tools, if not used properly, it can actually decrease productivity. A LAN can provide a means to streamline information processing and eliminate redundancies, but it may also deter users from logging on because of a perception problem. To the common user data communications and local area networks are a black hole of protocols, access schemes, routing algorithms, cabling and topology issues, and service problems. To alleviate these problems, the users should be educated about the basics of LAN technology and be provided with metrics and tools with which they can adequately wade through the myriad issues and select a LAN mapped to their needs.

When you look at the many options available for prospective LAN purchasers to evaluate, you can see the reasons for their distress. LANs can be very simple, providing simply an I/O channel and link to interconnect to another machine, or they can be highly elaborate, coming with their own distributed operating system, protocols, and services. The prospective LAN purchaser must decide what type of cabling is wanted or necessary, the type of electrical characteristics, signalling scheme, protocol for controlling transfers, routing schemes, topology of interconnection, reliability requirements, fault tolerance if necessary, services, interface characteristics and requirements, and numerous other aspects. The extent of control, understanding, and compatibility with other equipment a user wishes will decide which of these and other issues need to be addressed before a LAN is purchased.

In any case, addressing these issues requires performance metrics and evaluation criteria. In order to generate such information, a user must follow a methodology of selection that defines the user needs, the motivations, and the environmental and technological boundaries. As with the purchase of any product, the purchaser should identify how the product (in this case a LAN) will be used. This first part is the most important since if we don't define the need and uses properly, the remaining tasks will themselves have a predefined built-in error. Therefore, the prospective buyer should compile a wish list of all potential uses. For example, the list may include:

- Distributed file server
- Word processing
- Spreadsheet analysis
- Electronic mail
- Remote job entry
- Real time control
- Interactive log-on and execution or results
- Physical installation layouts
- Maximum node count and types
- Reliability considerations
- Network management
- Factory automation
- Computer types
- Video, audio, or both
- Interconnection to existing MAN or WAN
- Resource sharing
- Distributed computing
- Very large database

From this wish list the user must generate communications transfer and management requirements. For example, given that we have N computers which must be able to simultaneously transfer data to other sites, we have given a requirement for bandwidth (or an I/O rate maximum) and concurrency of access, both of which affect protocols, topology, and media requirements, to name a few. This set of communications transfer and management requirements can now be used to aid us in the other phases. The second portion of the methodology is to develop a motivational purpose for the LAN, to define why we want one in the first place. For example, we may want to compete with our competitors who are offering better or extended service to their customers by the use of a LAN, to have an edge in information availability to enhance the corporation's decision-making ability or to provide better control or use of the company's computing resources. The motivation for LAN selection will also provide our prospective buyer with more performance and evaluation criteria upon which to base a decision.

The next phase within the LAN evaluation methodology is to assess the environmental and technological aspects that the LAN must fit within. For example, is the LAN for a dirty, hot, cold, or varying environment? Will it be subjected to stress and strain from wind, rain, snow, lightening, and traffic? Will it be put in an air-conditioned computer room or strung through a building? Is the building new construction or old construction? Will it penetrate floors and go up risers? If so, what is the prevailing fire code? Will it link many buildings together? If so, will it be strung overhead or be poled from building to building? Will it be buried? Will it go under water or within water-carrying pipes?

From a technological viewpoint, the LAN may need to connect a diverse set of present company assets and also be able to link planned new resources. These resources have their own peculiarities in terms of electrical specifications, pin count, and makeup. These peculiarities will also map into requirements on the LAN. The LAN must be able to interconnect these devices directly or via an intermediate device, which should be an off-the-shelf component if possible.

Once all these initial analyses have been completed and their data compiled, the prospective purchasers should have a large volume of data from which to drive the LAN requirements.

The next question is: Is this data used to assist in the selection? Do you compile it into a model of a prospective system and use this to derive analytical and qualitative analysis of the prospective system and then compare these results to other known product parameters? Or is a simulation model more in line? In any case, a means of evaluating this data must be provided and it must be able to use data that has been collected.

The collected data can be divided into quantitative and qualitative classes. That is, there is one set from which specific performance measures can be derived and another from which only subjective measures can be derived. The quantitative ones should be used to build a model of the proposed system and derive composite measures to evaluate given prospective LANs. The methods used for this are analytical and simulation models. The test bed and operational analysis methods may not be viable to test alternatives on.

BOOK SUMMARY

The thrust of this book is to investigate the tools for performance evaluation of networks and provide a detailed look at one in particular: Malan.

In Chapter 2 the local area network is readdressed and looked at in further detail as to its hardware construction, topologies, control protocols, and systems management component technologies.

Chapter 3 readdresses the modeling issue from the slant of modeling local area networks, how the various tools have been useful in past systems, and how they can be applied to future endeavors.

Chapters 4 through 8 introduce further details about the applications, use, and technology associated with analytical modeling, simulation modeling, operating analysis, and test beds as applied to LANS.

Chapter 9 shows prospective purchasers the benefit of having a large volume of data from which to drive the LAN requirements. How is this data used to assist in the selection? Do you compile it into a model of a prospective system and use this to derive analytical and qualitative analysis of the prospective system and then compare these results to other known product parameters? Or is a simulation model more in line? In either case a means of evaluating this data must be provided, and it must be able to use data that has been collected.

This will be followed by an introduction and detailed look at Malan in Chapter 10. Following the review of Malan's structure, a detailed simulation model of LANs will be examined and results indicated. This will highlight the support that Malan offers to the LAN modeler.

2

Local Area Networks

What sets LANs apart from their early wide area network (WAN) ancestors are their variances in topology, connectivity, protocol, dispersion, communications rates, and communications equipment. LANs typically are high-speed shared media that interconnect devices in close proximity to each other (< 10 kilometers). The media is shared in the sense that multiple devices are strung from each link and there are dedicated channels between devices. The communications equipment that links the nodes together in a network is typically simpler (less costly) than for wide area networks. The Arpanet, for example, uses PDP-11 minicomputers as its imps on interconnection devices. They handle all the protocol for message transfers, whereas in many LANs the interface units now consist mainly of a few LSI/VLSI devices that implement one of the standard protocols (e.g., IEEE 802 standards).

The media of wide area networks has typically been satellite links or leased trunk lines, both of which are one-at-a-time media and are relatively expensive. However, in the local area networks, interconnection media such as twisted pair, coaxial cable, radio signals, and fiber optic cable have been the mainstays. All of these are much less expensive than their wide area network predecessors. Protocols are another area in which LANs and WANs differ. Wide area networks typically work in what is called a "store and forward protocol." In this protocol, the message to be sent is broken up into pieces called "packets." The packets are sent into the network using a routing protocol that will provide a means to get the packet to its final destination. The message is sent one packet at a time; it flows through the network via the routes selected by the routing protocols, arrives at the destination site (possibly out of order), is reassembled, and then is absorbed into the destination host. The LAN, on the other hand, uses one of three basic protocols to route the message to its destination. There are typically no intermediate steps, and the message is sent directly to the receiver. The reasons for the difference in protocol arise mostly from the differences in the topologies. The WAN (see Figure 2-1) uses point-to-point links to build up a network and must send all data from point to point in a hopping fashion; the LAN typically exists on a shared media (Figure 2-2). Messages on the LAN are not sent NIU to NIU in a hopping fashion. Instead,

18 Modeling and Analysis of Local Area Networks

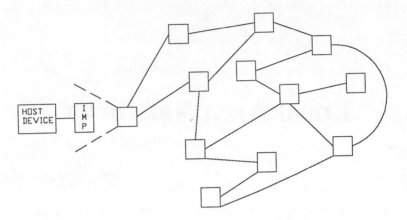

Figure 2-1 Wide area network (distance in hundreds of miles).

messages are broadcast over the network and all NIUs see it and examine it. The one to which the message is addressed accepts it; the others ignore it.

Topologies for WANs are typically referred to as irregular because there are no set symmetrical connections. The LANs are called regular or simple since the links are simple and have symmetry. The basic types of LANs are the ring, bus, star, and regular interconnects; they will be examined in greater detail later in this chapter.

From the above, some basic notions of what constitutes a local area network can be derived; namely:

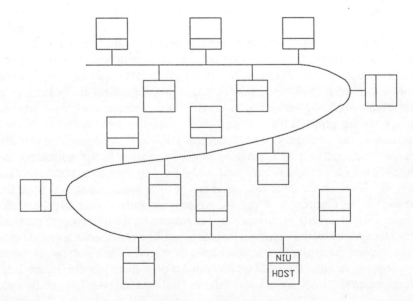

Figure 2-2 Local area network (distance less than 10 km).

1. Local area networks are used mainly to connect devices over short distances (< 10 km).
2. They typically are owned and used by a single organization.
3. Their topologies are simple and regular.
4. Their interconnection hardware and software are simple in comparison to those of WANs.
5. They use high-speed shared media.
6. Typically they interconnect homogeneous host devices.

LOCAL AREA NETWORKS: THEIR USES

Uses for LANs appear to be as numerous as computers, but they can be roughly classified into three areas: organizational, financial, and technical. The organizational uses deal mostly with the management issues of coordination, centralization, decentralization, and productivity enhancement.

Coordination implies the synchronization of activities to bring together dispersed elements into a unified action. Coordination is important to all organizations such as business, military, and educational. Without coordination an enterprise's actions could cause catastrophic failure rather than tremendous success; the traditional problem of the Byzantine generals will help to clarify what this aspect means. The generals are dispersed over some area. They are trying to decide the best course of action in their campaign, the goal being to all agree to attack or retreat. If a consensus is not reached, the generals will have failed. If, instead of using messengers who might die en route, the generals had a LAN to connect their camps (radio wave based), they could send messages (encoded) and get the proper answer using their fault tolerance algorithm. The LAN would provide them with greater communications abilities, thereby increasing their probability of successful coordination. Businesses, too, must coordinate buying and selling activities in order to be successful. The company with the most up-to-date information upon which to make its decisions typically will rise to the forefront. LANs provide businesses with the capability to communicate and coordinate business moves based on total knowledge of a corporation's status.

Another organization impetus to use LANs is to centralize control. This may seem to contradict the idea of networking. In the pure sense it does, but if the company's goal is to provide a corporate entity that can oversee, collect, and collate all of a company's diverse, distributed operations into a cohesive whole, a LAN may be the ticket. The LAN can be used to provide "computer central" with an online capability to extract and collect information in real time as opposed to having it called in or delivered occasionally. This maps back to the first notion of coordination: centralization is a means to coordinate. In a better sense, centralization provides a means for a company to build a status profile online and to use it to direct the satellites (other nodes on the LAN).

Organizations also go to the other extreme. They are too centralized and all decisions and data processing occur at a central site. The remote sites suffer in productivity because they cannot quickly (because of terminal and queuing delays, etc.) acquire access to, and use the data, at the central site to perform their tasks. They may also each have their own machine and require copies of all major components (hardware and software) to perform their job. If a LAN were provided, the dispersed organizations could easily distribute their information to all the other sites and also could access information and resources at the other sites. Providing a network and distributing functions from the central site to where they are mostly used lessens the load on the central site, improving its performance and at the same time providing better service to the remote sites by allocating processing and the necessary data closer to where it is needed. All this can be done without losing control of the corporation's information. It is all still online and accessible from any of the sites.

Productivity, the last organizational reason for using LANs, is less obvious, but it is derived from the discussions of the other three reasons. Productivity enhancement takes many forms, and a few are listed below:

1. Better information dispersal and integration
2. Resource sharing
3. Data sharing
4. Better ability to specialize computing sites
5. Better ability to tailor site needs
6. Greater availability and reliability of computing resources
7. Increased growth potential

Financial benefits can also be realized through the application of a LAN. A LAN can be used as a means of acquiring cost-effective computing. If an organization offers a link-up to its LAN with computing services available for purchase, users can realize a benefit in having great computing power at only the cost of a terminal, interface unit, and machine usage charge. This is much less than running your own machine. Such networks offer an array of devices from database research services to supercomputer usage. You use what you need, and the LAN makes it all possible. The LAN in this environment provides for a resource-sharing environment in which users log on and use whatever devices are needed to perform their function. The LAN need not perform this function, and it need not be part of a "pay for service" network; it could be just a corporate resource-sharing network providing all users with access to the corporation's large array of devices. From such a company's view the cost-related reasons for acquiring a LAN are:

- Consolidation
- Data exchange
- Backup
- Distributed processing

- Specialization
- Sale of services
- Resource sharing

Finally, there are technical reasons for LAN integration into an organization's computing center(s). The major technical drivers for migrating to LANs include:

- Faster turnaround
- Higher availability
- Expandability
- Load balancing

Technical reasons are always the ones pushed by the data processing center, which wishes to continually provide better service to its user community. Most centers find that some computers are busier than others in their center. The reasons may include "This is the one I'm hooked to," "I like the way this one services me," "I've always used this one," and "The software I want is on this system." The reasons go on and on. In any case, the computing center has a problem and can solve it by forcing users to be physically bound to another machine, by moving software, or by getting a common operating system. All of these, though, are temporary solutions. The center's assets sooner or later will hit boundaries. The multiplexer box for terminals may fill up, allowing for no more connections. The multiplexing devices may not be able to link multiple machines together, and if they can, the number of links is limited.

A LAN provides a solution to all of these problems. A LAN can be used to interconnect all the terminals and computers together into a single system. The LAN vendor may also offer fancy software to provide a uniform interface throughout, thereby alleviating another problem, and it provides a means to increase availability (all devices on line), expandability, increased turnaround (distributed processing, remote execution, etc.), and better overall utilization of assets (load balancing). In any case the benefits of local area networks far outweigh the potential drawbacks [Fortier, 1988]. The biggest issue is which one to choose and what items are to be considered in this choice. The next section will look at the last option, presenting an overview of some of the key components of LANs, namely:

- Topologies
- Interface units
- Control protocols
- Systems management

LAN TECHNOLOGY

This section will review some of the basic concepts of structure, operations and management.

Topology

Topology in a LAN or any other system refers to the pattern of interconnection for the associated components. This pattern can be very simple or rather elaborate; the sole deciding factor is the user's needs.

How the LAN is to be used and where it is to be placed provide the necessary data upon which to select a topology. What are the topologies available to prospective purchasers? The most common or basic topologies are the ring, bus, star, mesh, regular, and hybrids. These topologies and their varied combinations provide the majority of topologies found in present systems.

Ring

The ring interconnect in its simplest form consists of a set of nodes (interface units and host devices) interconnected from one to the other to form a ring. Each device has one input and one output link as shown in Figure 2-3. The elements connect their output to their neighbor's input which, in turn, does the same until the originator's input is connected to the last node's output. It is a true ring in terms of use in that if an NIU sends a message, it will see the message return on its input with a delay equal to a round trip time. The output of any NIU must always go to the same node (my neighbor) since it is physically connected to that device. In this

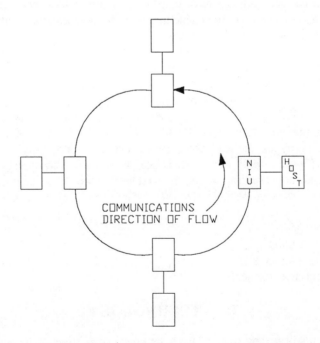

Figure 2-3 Simple ring interconnect model.

way, the ring structure causes messages to circulate around the ring from node to node in a circular fashion.

The protocols for communications will be discussed in later sections, but a quick description will aid in understanding this topology. The control of the flow (one direction only) is achieved through the use of a circulating ticket, or token. The token flows around the ring in one direction. When it arrives at an NIU, if it has any messages to send, the NIU appends them to the tail of the token train (sequence of circulating messages). When the train comes around again, the NIU removes the message it put on and copies any others addressed to it. In this fashion, messages flow on and off the network. This scheme can allow for many messages (up to the physical size limits of the ring length and clock rate) to flow simultaneously around the ring. Based on its simple architecture and operation, the ring topology has been used in many LANS, including the IEEE 802.5 token ring. This topology is attractive for the following reasons:

- Elimination of routing problems. All messages flow in the same direction.
- Expansion is easily accomplished by unplugging one connector, inserting a new NIU, and adding one additional link.
- Communications hardware is simple since all communications are point to point (from one output to one input over a dedicated link).
- Cost is proportional to the number of nodes. That is, n nodes require n links.
- Message throughputs can be high since no contention exists.
- Control as described is extremely simple (easy to use and has support in hardware or software).

However, the ring is not perfect: If a node or a link fails, the network fails (the ring is broken). To alleviate this requires additional links (bypass paths) and logic to detect and switch to the redundant paths. Alternatively, it requires multiple rings circulating in opposite directions with logic to detect breaks and cause loopbacks (transmission or alternative ring) to get around failed links (see Figure 2-4). With these added features the ring topology provides a simple system for interconnecting computers into a local area network. The advent of the 802.5 standard and the FDDI ring networks add to their availability and ultimate use.

Bus

The second basic type of LAN topology is the bus. Its basic structure is a single line with multiple devices tapped off of it (see Figure 2-5). The structure of this topology lends itself to using media that can be easily tapped into (T connections), such as twisted pair or coaxial cable. Fiber optic buses presently are not well suited to this use unless tightly coupled (close proximity) where a star coupler can be used to "model" the bus. The bus topology was one of the first used in local area networks. Probably one of the best known global bus topologies (and system) is Ethernet. This system uses the global bus to interconnect many devices with a contention-based control scheme (discussed later in this chapter).

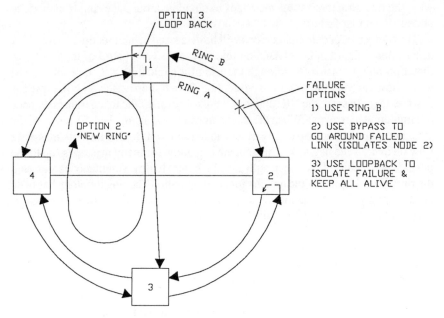

Figure 2-4 Dual ring with failures and bypassing.

Bus LANs have been designed and built that exhibit a wide range of capabilities and service. Typically, bus-structured LANs are used to interconnect personal computers, work stations, and other computers into "mid-range" performance networks; they are called this because, based on most protocols available for bus-structured LANs, somewhat less than 100 percent of the available bandwidth can be used. These types of LANs lend themselves nicely to interactive or bursty environments. The benefits of a bus-structured topology are:

1. It is simple to construct (single global line).
2. It provides a media for many protocols.
3. It is easily expandable (just add a node).

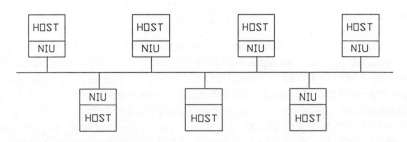

Figure 2-5 Global bus topology.

Local Area Networks 25

4. It provides good throughput response.
5. It has many standard interfaces available (802.3, 802.4).
6. It provides good fault tolerance.

Star

The third class of topology generally available for local area networks is the star. This topology is best described as a collection of nodes (NIU and hosts) connected together via a central "switch" (see Figure 2-6). All nodes in the network must be connected to this central switch in order to be part of the network and communicate with other nodes.

The central switch is an active device that accepts requests for communications and determines who to service next based on the protocol in use. This topology is the closest to a conventional timesharing system with its central host CPU and some number of terminals. Like the central timesharing system, the star topology is limited to the maximum number of points on the central switching device. Also like its timesharing predecessor, the entire network depends on a single device; if this fails, the entire network fails. This central device also poses another problem, that of performance limitation. The central switch must handle all traffic; therefore, it affects all traffic since as its queues of pending requests fill, the wait time associated with service will increase. The central switch is a bottleneck and will limit performance to some saturation point which it cannot go beyond. Additionally, because of the central switch limitations, the star configuration exhibits growth limitations, low availability, poor expandability, limited throughput, low reliability, and a complex central switch.

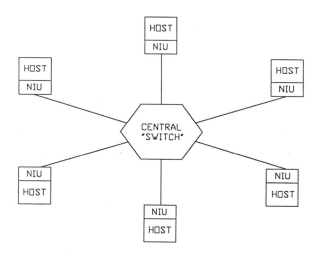

Figure 2-6 Star topology.

Mesh

The mesh, or irregular topology as it has been called, typically is defined as a collection of nodes connected in some random point-to-point fashion (see Figure 2-7). The topology consists of source and destination pairs that are connected in a random fashion similar to that of their wide area network ancestors. This topology has many possible variations and covers a wide range of complexity.

Mesh is one of the simplest topologies to implement in hardware, requiring that only point-to-point connections be supported (such as an I/O channel). Problems arise, however, in routing, flow control, congestion control, performance, reliability, and complexity. All of the problems and the solutions developed, postulated, and researched for the wide area networks of this type can be used. However, these solutions may not provide adequate throughput of messages at a LAN level. The delays associated with routing, packeting, flow control, etc., may be totally unacceptable for LANs.

Alternatively, if we use a more robust mesh (in terms of performance) such as the total interconnect (see Figure 2-8), we can solve some of the problems in performance and enhance reliability, but at a cost in connection complexity. The total interconnect requires a line for each node in the network to each other node. Therefore, a LAN of size n requires much more than the n links required in the simple mesh case. Additionally, as a node is added, we will need to add n - 1 additional links, but only if the other n - 1 sites can support an additional link. The growth of such a network is limited and it is expensive to implement.

REGULAR TOPOLOGY

Related to the irregular topology, or mesh, are the regular topologies. These are typically organized as blocks or cubes and have regular structures (numbers of

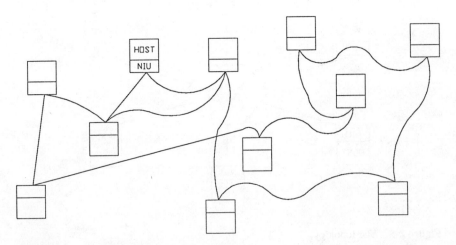

Figure 2-7   Mesh topology.

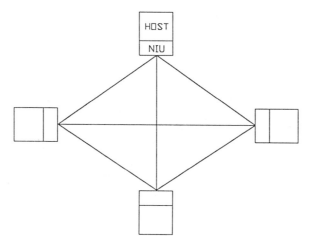

Figure 2-8 Total interconnect mesh.

interconnects) for all nodes (see Figure 2-9). They can also be arranged as shuffles and butterfly-type configurations.

The key to these types of LANS is that addressing (routing) is accomplished by presenting a bit string that represents the location of the intended destination on the network. Routing is simplified, but at a cost of providing hardware and software to do address translation and circuit switching. They potentially have all the same problems of the irregular networks (requiring less detailed routing) such as flow control, congestion control, etc., but they provide nice architectures for specific

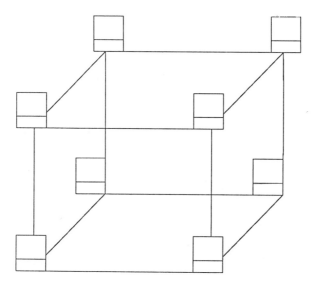

Figure 2-9 Regular cube topology.

28 Modeling and Analysis of Local Area Networks

problems such as data flow computing, systolic processing, etc. Details of some regular networks can be found in a special issue of IEEE Computer [Bhuyan, 1987] dedicated to interconnection networks.

HYBRIDS

Hybrid topologies are basically combinations of the previously presented topologies. We could, for example, construct a local area network as a collection of networks with all of the previously described networks used as a subset of the total (see Figure 2-10). In such a network, we could tailor the subsystem to jobs that could best use the type of topology associated with it. Such a network could provide very good service to properly tuned tasks, but incompatibility with information formats and protocols must be taken into account. If nothing else, it shows the many ways in which LANS can be used to construct novel, new systems.

An important issue in all of these topologies is that of media. The media selected for use will affect the selection of classes of interface designs and protocols in many ways. The selector of a LAN must keep this in mind when looking at topology.

INTERFACE UNITS

The topology of an interconnection scheme is but the beginning of a network. Regardless of the topology, all nodes must be connected into the network via an interface unit. Generically an interface unit contains five major elements or subsystems (Figure 2-11):

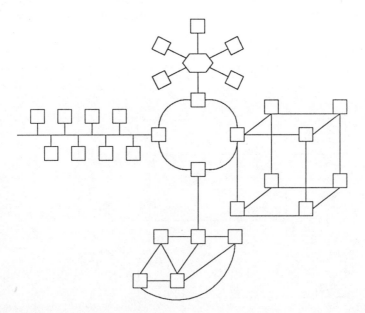

Figure 2-10   Hybrid LAN topology.

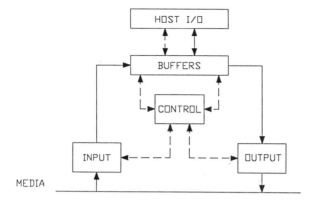

Figure 2-11 Generic interface unit.

- Media input
- Media output
- Interface control
- I/O buffers
- Host I/O

The input element extracts bits from the media, converting them to internal interface unit formats and delivering them to the buffers. Additionally, this device must be designed so that it detects a particular sequence of bits and, based on them, provides some further number of bits to the control element for further processing. It also strips off control words and performs error-detection checking.

The output device extracts the data from the buffers for transmission, formats this data for transmission, and transmits it over the media. As part of its formatting function it adds header information to bind the data into a message and supplies the necessary addressing and routing, flow control, and error detection information to the data.

The controller provides the "soul" for the interface unit. Its function is to direct how all the other elements perform their tasks. It indicates to the input controller which buffer to insert incoming messages into. It indicates to the input controller whether this message is to be pulled in or ignored (not addressed to it). It helps the input device, performing error detection and correction. The controller directs the efforts of the output device, indicating to it what buffer to read the message from, what destination to send it to, how to perform the error encoding, and what type of extra bits to add for the controller on the other end to use in its function. A third job of the controller is to manage the space of the buffers. In this capacity, it determines how many buffers to allocate for input, how many for output, and which messages get associated with what buffers. Additionally, from the host input-output controllers the control unit helps to accept and decode access to read and write messages to and from buffers and to map user addresses to buffers to network addresses.

In its capacity for protocol control the interface unit controller must provide services for address translation and mapping, segmentation, packetization, security, routing, and fault detection as well as basic protocol functions of network control (who to put next).

Interface units come in all sizes and complexities. For instance, interface units themselves can be whole computers (microprocessor and hardware interfaces), they can be built from discrete logic (TTL, LSI, etc.) of varying complexity, or they can be VLSI implementations of protocols; however, in all cases the basic components of the design don't change—just the implementation does. In some cases, the logic associated with providing the network interface unit functions can itself represent more hardware and software then the host it is meant to support. This will be illustrated in Chapter 8 when a test bed implementation of a network is reviewed.

On the brighter side, technology has been constantly improving NIU designs, especially since the development and release of the IEEE 802 standards. Because of these standards we can now readily purchase chip sets to provide token bus, contention bus, and token ring networks (see Table 2-1 for protocol type and manufacturer). The interface units provide the means to implement the protocols of control and to manage the flow of information over the network.

CONTROL PROTOCOLS

Protocols are used in all aspects of life. They provide us with the means by which we are to interact with one another. For example, in the world of politics protocols exist to describe how countries are to act in various situations. They are also used in business to describe how one is to act toward other people. In the communications world and in particular LANs, protocols describe and control how the devices or the networks are to communicate with each other.

LAN protocols exist as layers, each related to performing controlled communications with a particular hardware or software layer (see Figure 2-12). These protocols provide the means by which we examine link bit patterns and determine what to do in order to effectively and correctly communicate information over the communications media. There are three basic classes of communications control protocols: contention based, reservation based, and sequential based.

Contention-based protocols operate by having the interface units that have a message to send actively competing (by various mechanisms) for control of the physical media. The reservation-based protocols operate by the slotting (logically) of the media communications' capacity into pieces and the allocation of these pieces among the devices on the networks. The sequential-based protocols operate the rotation of the media access in a controlled fashion from unit to unit. In all cases, however, the basic premise is to provide a means to unambiguously allocate the media to an interface unit so that it may communicate information to another device in the network.

Table 2-1 LAN Standard IC-Chip Sets

Standard	Protocol	Manufacturer	Chips
802.3	CSMA/CD EXAR XRT	Advanced Micro Devices	7996, 7990 820515, XRT 820516
	INTEL		82586, 82501, 82588
		National Semiconductor	DP8790, DP8341, DP8342
		SeeQ Technology	8003, 8023
		Thomson Mostec	MK 68590, MK 68591
		Western Digital	WD 83C510
802.4	Token passing	Motorola	MC 68184, MC 68824
		Signetics	NE 5080, NE 5081
802.5		Texas Instruments	TM 538030, TM 538010, TM 538020, TM 538651, TM 539052

CONTENTION BASED

As noted briefly above, contention-based control protocols are based on the interaction of interface units with each other to acquire the network media for service. Two major classes of service type exist under this classification: carrier sense multiple access (CSMA) and request-grant access (RGA).

LAYER-7	APPLICATION
LAYER-6	PRESENTATION
LAYER-5	SESSION
LAYER-4	TRANSPORT
LAYER-3	NETWORK
LAYER-2	DATA LINK
LAYER-1	PHYSICAL LINK

Figure 2-12 Open standard interconnect protocol layer model.

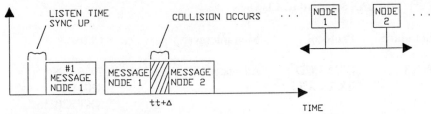

Figure 2-13 CSMA collision.

In the CSMA mode of contention, user devices that wish to communicate to some other device simply begin transmission on some sequencing boundary (in the basic case this is simply a wait of some idle time on the media). When the idle time (wait time, synchronization time, etc.) is met, the devices actively contend by sending out their messages. At the same time that they are sending, they listen to the media to see that what they are sending out is the same as what they are hearing. If it is after some time period (typically twice the round trip bit transfer time), they have acquired full control. If it does not match (because two or more devices are sending at the same time), a collision has occurred and the sender will immediately cease transmission and wait before sending again. The variation on this protocol deals mainly with how senders wait (if at all) and how they initiate and detect a problem.

This example of operations will help the discussion (see also Figure 2-13):

- Node 1 wishes to send two messages over the network.
- It sends its message out and starts to listen.
- The message goes out with no problem.
- Message 2 is begun.
- Node 1 listens to the network.
- The message goes out with no problems until time t.
- At time t node 2 begins its transfer.

By time t both nodes know there is a collision because the messages are garbled. That is, what was sent is not what is being received.

In pure collision technique, the message would be sent out in total; if collisions occurred, they would be resent at some later time.

In CSMA schemes as soon as the collision is detected, both nodes stop transferring information. To deal with the collision, the CSMA techniques retransmit the message based on some criteria, such as:

- Persistent
- P-persistent
- Nonpersistent

The CSMA techniques use time period a to determine if a collision has occurred. This a is equal to a round trip time delay (the time it takes a bit to transfer from the two most extreme portions of the network and back). The collision detection schemes rely on the use of this knowledge to aid in detecting errors and for sensing if the bus is busy.

The key aspect is the sensing of the activity on the bus. In the persistent case a node that wishes to transmit senses the bus; if it is busy, it waits and continues to listen. When it becomes free, it waits 2a, or one round trip time delay, and if still free, it begins to transmit its data. It then continues to sense the bus; if a collision occurs, it halts transmission immediately and waits for another 2a free time and tries again. This continues, with an increased delay inserted based on the number of times a collision has occurred.

In p-persistent the protocol operates in the same way to detect a free time, but the difference now is that the node will transmit its message based on a probability p. If it chooses correctly, it will send; otherwise it waits with the probability 1 - p. This technique provides for additional separation of colliding devices, but it may cause undue delays at times.

The last form of CSMA protocol is the nonpersistent. In this form, the contenders, upon having a collision, will not try again immediately as in the other two cases. Instead, the protocol chooses some random time delay and goes to sleep until this time. When it awakens, it reenters active contention for the network resources. Schwartz [1977] has a nice analysis of this technique in his book on queuing systems as applied to networks.

The second class of contention-based control protocols is request-grant access (RGA). In the basic form of this contention scheme, the units that wish to acquire service request it and, via a resolution algorithm, either succeed or fail in their attempts. The algorithm can be centralized or distributed based on the media being used and the topology. For example, a star network can very easily use a simple request-grant scheme in which all the devices ask for service and the central hub grants it based on some ingrained priority.

A novel form of request-grant contention access is found in the dot-or protocol [Fortier, 1984]. In this protocol the requesters contend by sending out their message priority followed by their physical device address in a bit-serial fashion. As the bits are transferred, they listen to what is being received; if it does not match what they sent, they drop out of contention. If it does match, they continue on to the next bit, and so on. To break the inevitable message priority tie, the network uses the physical device address. If a message tie occurs, the contention continues through the node address dot array. At this point, since the devices are given a unique system-wide address, a winner (the one with the larger ID number) will be found and the bus assets will be allocated. The scheme looks to schedule the next "frame" (time slot) and works by reserving the next time slot for the winner to send data over. Details of this can be found in Fortier [1985] and its references.

RESERVATION BASED

In reservation-based control protocol schemes the interface units win use of the media based on preallocated slots of time. The allocation of slots can be equal or it can be unequal; this is based on what style of service is required.

There are two types of control protocols typical of this scheme: slotted and vectored. In the slotted scheme the communication media assets (time) are broken up into slots. These slots can be fixed in size or can vary. The problem is how to preallocate them in a fashion that will be fair and will provide adequate service to users. A well-known type of preallocation is that applied in the Honeywell experimental distributed processing (HXDP) system. In this architecture and others like it, control of the media is done via a simple allocation list. The list has available slots and those that this node controls. In the HXDP case the list is a 256-bit rotating buffer. If the pointed to bit is a 1, then control is granted; otherwise, it is not. This scheme works well in the distributed case since no global data must be distributed once the list is set up. The problems arise if there is a failure and two or more nodes go on at once. In this case the messages will be garbled and they will be lost. The control is affected by the signaling of completion by the last unit, at which time it signals all to update their pointer (see Figure 2-14). The pointer points to the same location in all the allocation lists. If a node has a slot reserved, it gets it; otherwise the node waits for the next update and tries again.

Another form of reservation-based control protocol is the vectored protocol. In this class of protocol the unit having control has a vector list. This list indicates which unit is to receive control next (an address) (see Figure 2-15). The pointers to the location in the list are not controlled in unison. The update of a pointer occurs only when a node has control of the control token (unique pattern). When a node acquires control, by being addressed by and given the token from the previous controlling node, it uses the bus; then after sending out the token to the node pointed to in the token (node n, for example, in node 1's table), it moves the pointer to the next unit that will receive control (for node 1 in the Figure 2-15 it is now node 2 at pointer +1).

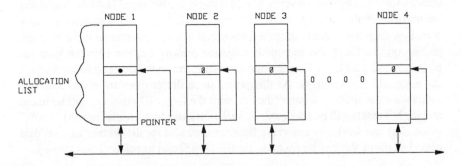

Figure 2-14 Reservation-based control protocol.

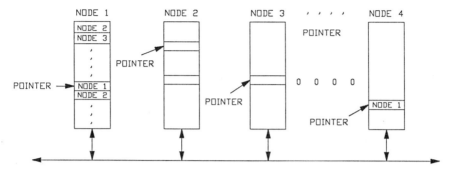

Figure 2-15 Vectored control protocol.

The node that was previously pointed to has been allocated the media and can use it or not. If it does not use it, the node reissues the token to its addressed unit and increments its pointer. In this fashion, many varied distribution patterns are possible. The problem is that it may not be giving control to the unit that needs it most; it may in fact be allocating it to units that do not need it at all. The problem to be solved by the designers is how to allocate the slots. This scheme is simple and requires little logic other than a circulating buffer and a list of physical addresses in the network and a means to read and write them from or to the network.

SEQUENTIAL BASED

The third type of control protocol is the sequential based, in which control flows from unit to unit in a logical or physical sequence. No repetition of nodes is allowed until all have been "slapped" (accessed) once in the sequence. The simplest form to visualize for this scheme is the token ring allocation scheme [Farber, 1975].

In this scheme control is given by the reception of a control token that is passed from node to node. For example, in Figure 2-16 the control will pass from node 1 to node 2 when node 1 passes the token from itself to node 2. When node 2 receives the token, it sends out its messages since it now has control. When node 2 completes its transmission, it sends the token out to node 3 and this node sends it to node 4 when it is finished. The sequence ends or restarts when node 1 receives the token from node 4. At this point a single scan of the network has occurred with all units receiving control once during the sequence. The major difference in this mode of control versus the others is that it must allocate the resource (network) to all nodes once before it starts again. In the others any single node can receive the right to use the media as many times as it is designed to.

In contention-based protocol, access is limited to how many times a message does not collide with another message. In the reservation-based protocol it is dependent on how many allocation slots a node has. A variant of this scheme does not use the physical sequence (as in a ring) to maintain the sequential operation,

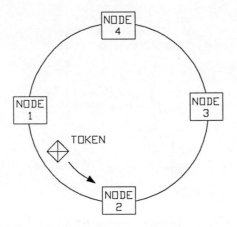

Figure 2-16 Sequential control protocol.

but a logical sequence. The system is organized any way we wish with the nodes knowing who follows them (see Figure 2-17). The allocation of the bus follows by issuing the address in the allocation register (pointer). This allocation is fixed and guarantees the sequencing of the token from node to node and back again with all nodes receiving access once per scan.

The issues in all these techniques are based on usage patterns and styles. The LAN purchaser who wishes to select an optimal topology and make it work for a specific application must also choose an appropriate control protocol that will deliver the type of service necessary. If a sequential protocol is chosen and the use of the network is very bursty, poor utilization will result. In this case, a contention-based node control would better serve such usage. On the other hand, if all nodes have messages to send, a sequential-based control protocol that gives equal service

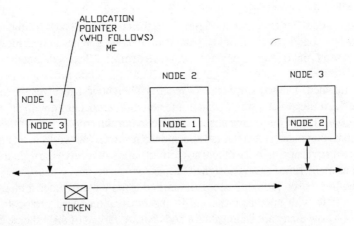

Figure 2-17 Logical sequence.

to all may be best. Finally, if there is a very fixed environment in which communications needs are well-known and described, a reservation scheme may be best. In any case, the control scheme is a critical component of selection for a potential purchaser to consider.

SYSTEMS MANAGEMENT

As important as communications are to the LAN's usefulness, so are the network systems' management components. Without low-level support to perform transfer of data, there would be no networks. Likewise, without good management components there would be no usable networks; that is, without excessive user invention a system could not be used effectively to perform some end-user function.

The system management components can provide simple services (network use) or very integral servers (a distributed operating system that provides for the management of network-wide process, memory, I/O, devices, the network, and files, to name a few).

The addressing of these diverse issues is beyond the scope of this introduction to the topic; refer to [Fortier, 1986, 1988, and 1989] for details. The major components of systems management to be addressed here are:

- Operating systems
- Database managers
- Network manager
- Error management

There are three flavors of operating systems for networks: the bland (no network operating system at all), the lightly seasoned (network operating system), and the full-flavored (distributed operating systems). Many LANs are purchased and come with no operating system support. The user is told how to interface to the LAN and must design and develop any support code necessary to make the job easier. The user has no aids in network addressing (in terms of processes, etc.). All the functions of an operating system must be handed by the user's code.

The second class of operating system is the network operating system, which is network support software that is added on top of an existing operating system (see Figure 2-18). A network operating system provides users with the ability to "use" the network. For example, if users wish to schedule a process to run elsewhere because of the beneficial aspects of the remote site (vector processor, database process, AI machine, etc.), they must know they are going into a network, that they must log onto the network, associate their process with a remote server, and request that service be performed. The remote server will then insert the process into the local operating environment, await completion, and signal results to the originator. If the two processes needed to synchronize activity, they would each need to be associated with a network operation system (NOS) server process that would send

38 Modeling and Analysis of Local Area Networks

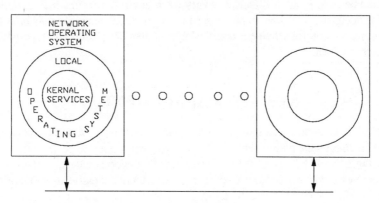

Figure 2-18 Network operating system hierarchy.

messages between them. They in turn would need to know how to interact with the server to perform the synchronization. The important aspect to gather from this brief overview is that a NOS is only a service added on top of an existing operating system that provides network control. It provides a means to simplify the use of the network through a set of services.

The distributed operating system deals with the network and local nodes differently (see Figure 2-19). In the case of a distributed operating system, all resources and user processes are under its control. If, for example, any site has a process it wishes to schedule, it provides this to the distributed operating system as a regular process (that is, it does not have to know a network is involved or name a node to place it on). The operating system will examine the process control block to determine the process-specific requirements (resources required, etc.). It then uses a global scheduler to determine where the best place in the network to service this process is located and to schedule it. The emphasis is on the global nature of operating system functionality and goals. The distributed operating system in this case is the native and only operating system for this network. Details of operating systems of this class and their technology can be found in [Fortier, 1986].

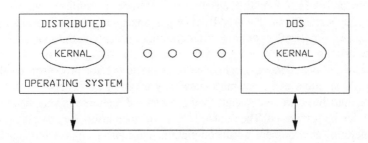

Figure 2-19 DOS view.

DATABASE MANAGER

Another important feature for systems management within a network to possess is data management. As was the case in operating systems, we can have varying degrees of support for data management from none to a fully distributed database manager. The basic question to ask from a potential purchaser's view is: What do I need and how does this map to available systems? If your goal is to have a file system, database managers are not a concern. But on the other hand, if you are performing some critical task requiring synchronized data at all sites, a fully distributed database manager is exactly what you need.

The levels of database management served by LANs include:

- No support
- File transfer
- Remote file management
- Local data managers
- Remote service to local data manager
- Integrated distributed database manager

With no support the LAN users must design, develop, and implement their own database manager and all its associated components. This option is fine if no support for secondary storage management is needed. Simple local file management is typically available as part of a user's own operating system even on very small PCs. In a NOS-type environment there are typically some services, in particular remote file servers, and management for such remotes is typically provided. This allows for the storage and use of files on remote sites, but it is under the control of the user. The remote file server provides some rudimentary services for storage and management of the remote file directories, providing a more systematic control.

An even newer aspect is LANs with remote databases and database managers. Typical of these are PCs with local data managers and possibly remote NOS-based servers to formulate and send database queries to directed sites. The user still has the burden of knowing what data is stored where and potentially even knowing the nuances of the particular sites' database access languages. A more recent advance has been the development and marketing of totally integrated distributed database management systems (DDBMSs). Those systems provide for a network-wide data manipulation language, data directory and dictionary support, and database manipulation optimized on a global network level. This type of distributed database system is usually only found if a distributed operating system is available. This is because the DDBMS requires the global support of access and computation synchronization to provide the beneficial features of DDBMSs, such as concurring control, consistency, correctness, and optimality of data in the system. Details of distributed databases and their benefits can be found in [Stone, 1980].

The third component of systems management is network management. From a high-level network management view the goal of this component is to provide transparent use of the network to access devices, resources, and other processes based on logical (human) names versus specific machine names. This component must also provide mechanisms for synchronization, multiple concurrent accesses, monitoring, routing, flow control, and fault tolerance.

The most important feature of network management is name management in terms of binding logical addresses to physical addresses and the management of these on a global level. All elements of the system must be allowed to use names and identifiers devoid of any network connotation. This requirement is paramount to providing a resilient and responsive system for user and systems processes and resources. Details of this type of component can be found in [Fortier, 1986 and 1988] and their references.

The last major component of systems management is the error manager. Error management covers a wide spectrum of issues. For example, error management can deal with low-level detection and correction of media errors or with end-to-end driver-to-receiver errors. They could also address errors as a user end-to-end problem or anything in between. On the micro level error management deals with mechanisms to detect errors (such as CRC, parity, etc.) on the media, to isolate where the error has come from, and to develop a plan to correct and recover from such errors. On the upper level error management looks at overall computations and their correctness, developing schemes to detect an invalid computation, to isolate where it came from, and to determine what it affected. From this the error manager must develop a scheme to correct or recover from the error. This may include performance monitoring components, fact isolation components, and reconfiguration or fault tolerance hardware or software to recover and continue operation in the face of errors. Details of these concepts can be found in [Fortier, 1986, 1988] and their references.

SUMMARY

This chapter provided a framework from which we can go on to describe performance evaluation of local area networks. It provided a description of what constitutes a local area network, how it is used, and for what purposes, and gave details of LAN topologies (interconnectivity). Included in this was an examination of the basic topologies: ring, bus, star, mesh, regular, and hybrid. This discussion led into an introduction to network interface unit architecture where the reader was led through the components of input, output, control, buffers, and host I/O. This discussion focused on the generic theory of operations rather than a detailed review of an architecture. The next section introduced control protocols and reasons why one is superior to another and in what cases. The contention-based, reservation-based, and sequential-based protocols were reviewed. This review was followed by a quick investigation of system management issues for LANs and how these can affect a potential purchaser's choice of a LAN.

REFERENCES

Farber, D.J. "A Ring Network," *Datamation*, February 1975, pp. 44–46.

Fortier, Paul J. "A Communications Environment for REAL Time Distributed Control Systems," Proceedings of ACM Northeast Regional Conference, 1984.

Fortier, Paul J. "Design and Analysis of Distributed REAL Time Systems," McGraw-Hill, 1986.

Fortier, Paul J. "Design of Distributed Operating Systems," McGraw-Hill, 1986.

Fortier, Paul J. *Handbook of LAN Technology*, McGraw-Hill, 1989.

Schwartz, "Computer Communication Network Design and Analysis," Prentice-Hall, 1977.

Stone, "Introduction to Computer Architecture," SRA, Inc., 1980.

3

Modeling of Local Area Networks

The previous chapters addressed modeling as one issue and local area networks as another. This chapter brings them together to present a view of how modeling can aid in LAN selection, design, and research.

Local area networks represent an important component of a user's computer system resource. As was said earlier, it provides the means to interconnect and unify disjointed computing resources into a computer network. The selection of a LAN to best meet the needs of an end user is a critical component in the system selection process. Whether it is a completely new system or an integration of a network into existing equipment, the process of selecting a LAN is the same.

The potential LAN purchaser must develop a set of requirements, perform some type of selection process (modeling, document review, analysis of existing implementations, or all of the above), select a LAN, install it, and then put it into use. The selection process typically requires a performance evaluation of the network versus some criteria such as speed, throughput, overhead, workload mix, number of nodes supported, etc. If proper performance evaluation and prediction is performed, the selection process should be much more fulfilling.

Good performance evaluation and prediction must begin early in the product selection phase and the selection criteria must be developed and documented early in the process. That is, the applications to be performed and the expected workloads to be applied to it must be defined as must the services required of the prospective system (hardware, software, network). The applications refer to the type or class of the end user process (e.g., mail services, database services, remote operating, remote process execution, distributed computations, real time control, data processing, etc.), whereas the workload implies the volume or amount of service requirements placed on the system based on the applications.

The second major consideration when looking at a LAN or making any computer selection is the hardware and software. Once we know what it is that the end system is intended for, we must characterize what it needs to support this and how much power versus cost we are willing to trade off. Typical of this phase is determination

43

of the number and types of hosts, specialized equipment needs, and storage and support resources. These decisions form the architectural boundaries of the system and define the source of its support. These criteria aid in the determination of metrics upon which we can model a system's (networks) performance to aid in the selection of an end product. To perform such a selection the organization must construct models of the intended system's architecture. These models can then be used to predict the performance of the system as if the actual new components (network) were available. Such analysis provides invaluable data to support the selection of an optimal LAN based on the modeler's knowledge.

Modeling LANs can provide the modeler with insight into the limitations of a proposed network before it is purchased, define bottlenecks in its architecture, aid in the selection of alternatives, or indicate ways in which to configure a proposed system to optimize its performance. In addition, all of these can be determined before money is expended to purchase and install a LAN, potentially saving much capital and time. Using modeling techniques, the prospective purchaser can make optimal (or nearly so) choices, thereby providing positive benefits of LAN technology and use versus the problems possible in a bad LAN choice.

In order to make such LAN models useful we must investigate them at a reasonable level of detail. But what is reasonable? Do we need to model the prospective LANs from a simple queuing representation (see Figure 3-1)? In this case our analysis of the LAN's potential as a communications server is limited to our ability to represent its service as a service distribution and its control protocol as a queuing discipline. In some cases this will suffice, while in others this will be wholly inadequate. The alternative to modeling at such a high level is to model at lower levels of detail, but what does this imply? Is a lower level to include simply multiple queues, representing the nodes and a more detailed model of the network? Or do we simply model all components as close to reality as possible? The real

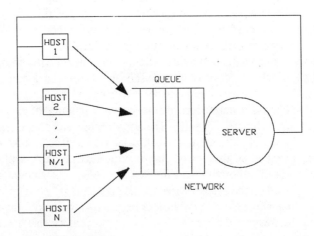

Figure 3-1 Simple network model.

decision lies in defining what level of detail is sufficient to answer the questions we wish to ask of the network.

In the extreme case we may model a network down to the transfer of single bits. With this type of model we can examine the lowest levels of a design. However, to perform such modeling will cost much more in terms of expense in developing, debugging, testing, using, and analyzing results. A better alternative if this level of examination is required is to construct a test bed or perform operational analysis on some other user's present setup. Low-level modeling such as was just described can give us enormous insight into how something operates and how it will react in many situations. This form of modeling has its merits but only when detailed use of the particulars of a product is to occur. Realistically, we will choose something in the middle. That is, we will use the highest and lowest levels of the system's complexity to determine where within the spectrum to perform the modeling (see Figure 3-2).

The two metrics we must measure against are depth of model (closeness to reality) versus its complexity. For example, in Figure 3-2, we may wish to faithfully model half of the components of a network but not wish to pay for a high complexity. If this is the case, particular items of the network that add excessive complexity to the model and add little to the outcome (such as signaling type) can be neglected, leaving detailed modeling to the items of most importance (e.g., control protocol, topology, communications, and throughput). This represents a trade-off in wellness of fit of the model and, therefore, its results versus the complexity and cost must be considered. The proper balance of detail at varying levels will supply the best model based on its intended use.

For example, if we wish to model a network control protocol and get data on network utilization and host use patterns, we can lower the detail on the control protocol (who gets served next) and increase abstractness at the other layers (e.g., the network transfer mechanism and NIU device physical transfer details), leaving the heavy modeling details to the control protocol only. As will be seen in later

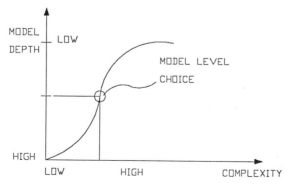

Figure 3-2 Spectrum of models.

chapters, this aspect is not a trivial matter and must be given early emphasis in model development. If it is not, the modeling effort may not reflect the correct reality that the modeler intended and it may not be able to provide answers to the intended questions.

Related to this aspect of level of detail is that of quality of model. If we choose a level of detail that is sufficient to provide the answers but do not develop a model that consistently uses this level, the model may be useless since it will not provide consistent results as was originally intended. This is as much a philosophical model as a technical one. Technically speaking, if the model's level of detail is adhered to, its quality of match to the original is consistent.

Quality in these terms is hard to assess in the sense that we cannot see a flaw with it until it has caused a problem in the performance analysis and prediction of an end system. Quality is a nebulous entity that is more qualitatively determined than quantitative; that is, quantitative terms provide us with indications as to quality but do not define quality. Quality is in the eyes of the beholder in terms of fit to reality. In one modeler's eyes a model may have high quality because it faithfully models a real world system in some aspect. However, another modeler may deem this model totally inadequate because it only models this one aspect faithfully.

What this discussion illustrates is the modeler's dilemma. A quality model will meet all verification and validation requirements where a low-quality model will not. In any case, the above discussions point to the validity and usefulness of modeling LANs to aid in performance assessment and therefore selection. Models provide the LAN purchaser with a tool to use in determining the best fit for an ultimate system's auxiliary storage, I/O channels, network interface units, and the network media itself. Jobs or programs in the system demand service from these resources or servers. The major performance problem that we can study with such a modeling tool is that of contention for service among the varying jobs, programs, and resources. Typical measures deal with delays associated with the related queues. The theory of queuing analysis provides a framework based on sound mathematical formulations to construct and solve problems related to this customer-server problem. Typical measures for a local area network that we can determine from such models include:

- Total time in system for all messages
- Throughput of links
- Utilization of links
- Mean response time
- Standard deviation
- Effect of various queuing disciplines
- Effect of various service disciplines
- Round trip time (message transfer) minimum, maximum, and average
- Effect on above based on queue interactions

The above represent a few of the potential measures and uses of analytical models for evaluating. They provide for:

- Controlled experimentation
- Realistic time compression and expansion
- Sensitivity analysis
- Presystem assessment
- Component performance and assessment
- Environmental interactions
- Bottleneck analysis
- Configuration tools
- Performance analysis and assessment

TECHNIQUES

As was described in the previous chapters, there are many tools that can be applied to the modeling of a local area network: analytical modeling, simulation, test beds, operational analysis, and hybrids.

Analytical Models

Analytical models can be used to study the steady state performance of a LAN in terms of performance criteria that can be expressed in analytical equations. The one that best lends itself to use in LAN evaluation is queuing analysis. In this class of analysis the computer system is viewed as a multiple resource system in which the resources are the CPU, memories, and local area networks. The following chapters will provide much more detail about their use from a purely theoretical, as well as practical, sense.

Analytical models provide a quick and relatively easy means to assess boundaries of a system. This is true as long as the analysis is limited to elements that can be described by analytical means. The problem arises when an attempt is made to use the analytical model to fit a problem of unmanageable size. That is, as the complexity of the system being modeled goes up, the analytical model's tractability goes down. There is a limit at which analytical tools cannot be applied because of attendant complexities. The alternative in such cases is to revert to another more flexible modeling tool, simulation.

Simulation

Simulations provide us with a means to model the local area network to any level of detail we deem necessary to effectively perform the wanted analysis. Simulation includes a wide spectrum of techniques such as the queuing of discrete events and continuous or hybrid simulations. Each of these techniques provides for various levels of detail; the choice is up to the modeler. Details of simulation techniques, models, and languages will be covered in the chapters to come.

Test Beds

Test beds are highly useful tools in the selection process for a local area network. They provide a laboratory environment in which researchers can perform a variety of controlled experiments that provide myriad data for purchasers to use in selecting a LAN. The issue in test beds is not in their usefulness but in determining where such systems become cost effective. A test bed is not cost effective if a user is looking at installing a small number of devices into a network. The LAN purchase, the development of test software and hardware, and the time associated with all these activities far outweighs the benefits of the test bed. On the opposite side of the coin, though, if we are looking at constructing a large LAN such as Southwestern Bell in St. Louis where more than 2000 workstations are connected over 90 miles of cable, a test bed to review alternatives is highly cost effective. The important feature of a test bed is that it provides a piece of the actual system that can be tested, stressed, and analyzed from all angles in a very controlled fashion. For large developments, such a capability is invaluable.

Operational Analysis

Operational analysis has also proven itself as a useful evaluation tool for LAN selection. LANs have been around for quite some time, and just about every LAN operation maintains performance measuring, monitoring, and assessment capabilities. The data they produce can be used by a prospective buyer to ascertain the performance (postulated) that this LAN can provide in the buyer's environment.

Operational analysis has given us many insights into network performance for many years. The early systems such as Arpanet were wired for monitoring, measuring, and analysis from the start. As such, these networks provided us with many measures of effectiveness and criteria to use in determining the performance of a network. Other early LANs such as Aloha Net and Ethernet have considerable data available on their operations in actual systems. This type of operational data is invaluable in assessing the potential for a similar network to provide any organization with adequate services.

SUMMARY

This chapter outlined some of the issues and uses of modeling in the LAN environment. It showed the usefulness of modeling as a tool to aid in LAN selection as well as to show the use over the life of a network. Covered were aspects of determining the usefulness of models, the level of detail available or necessary, and the quality that can be expected from various models. This was followed by a quick reassessment of the basic models and their uses in LAN evaluation.

4

Probability and Statistics

Queuing theory and queuing analysis are based upon the use of probability theory and the concept of random variables. In this section, we will introduce the concepts of probability and random variables and conditioned probability, followed by a section on random variables. We will then move on to probability distributions, stochastic processes, and finally, into the basics of queuing theory.

Before discussing queuing analysis, it is necessary to introduce some concepts from probability theory and statistics. In basic probability theory, we start with the ideas of random events and sample space. Take, for instance, the experiment that involves tossing a fair die (an experiment typically defines a procedure that yields a simple outcome that may be assigned a probability of occurrence). The sample space of an experiment is simply the set of all possible outcomes, in this case the set {1,2,3,4,5,6}. An event is defined as a subset of a sample space and may consist of none, one, or more of the sample space elements. In the die experiment, an event may be the occurrence of a 2, or the event that the number that appears is odd. The sample space, then, contains all of the individual outcomes of an experiment. For the previous statement to hold, it is necessary that all possible outcomes of an experiment are known.

Several operations on the events in the sample space yield important properties of events. By definition, the intersection of two events is the set that contains all elements common to both events. By extension, the intersection of several events contains those elements common to each event. Two events are said to be mutually exclusive if their intersection yields the null set. The union of two events yields the set of all of the elements that are in either event or in both events.

The complement of an event, denoted $\overline{A}$, represents all elements except those defined in the event A. The following definition, known as "DeMorgan's Law," is useful for relating the complements of two events.

$$\overline{AB} = \overline{A} + \overline{B} \tag{4-1}$$

Permutations and combinations of elements in a sample space may take many different forms. Often, we can form probability measures about combinations of sample points, and the basic combinations and permutations discussed below are of use in this task. By definition, a combination is an unordered selection of items, whereas a permutation is an ordered selection of items. The most basic combination involves the occurrence of one of n_1 events, followed by one of n_2 events, and so on to one of n_k events. Thus, for each path taken to get to the last event k, there are n_k possible choices. Backing up one level, there were n_{k-1} choices at that level, thereby yielding $n_{k-1}n$ choices for the last two events. Following similar logic back up to the first level yields

$$n_1 n_2 \ldots n_{k-1} n_k = \prod_{i=i}^{k} n_i \tag{4-2}$$

possible paths. For example, in a string of five digits, each of which may take on the values 0 through 9, there are $10 \times 10 \times 10 \times 10 \times 10 = 100{,}000$ possible combinations.

When dealing with unique elements that may be arranged in different ways, we speak of permutations. For n distinct objects, if we choose one and place it aside, we then have $n - 1$ left to choose from. Repeating the exercise leaves $n - 2$ to choose from, and so on down to 1. The number of different permutations of these n elements is the number of choices you can make at each step in the select and put aside process and is equal to

$$p(n,n) = (n)(n-1)(n-2)\ldots(2)(1) = n! \tag{4-3}$$

The common notation $P(n,k)$ denotes the number of permutations of n items taken k at a time.

Choosing from n distinct items taken in groups of k at a time yields the following number of permutations.

$$p(n,k) = (n)(n-1)(n-2)\ldots(n-k-1) = \frac{n!}{(n-k)!} \tag{4-4}$$

The permutation yields the number of possible distinct groups of k items when picked from a pool of n. We can see that this is the more general case of the previous expression and reduces to Equation (4 - 3) when $k = n$ (also, by definition, $0! = 1$).

The number of combinations of n items taken k at a time [denoted $C(n,k)$] is equivalent to $P(n,k)$ reduced by the total number of k element groups that have the same elements but in different orders [e.g., $P(k,k)$]. This is intuitively correct because order is unimportant for a combination and, hence, there will be fewer unique combinations than permutations for any given set of k items. $P(k,k)$ is given in (4-3), hence

$$C(n,k) = \frac{n!}{k!(n-k)!} \tag{4-5}$$

Alternatively, one can state that each set of k elements can form $P(k,k) = k!$ permutations which, when multiplied by $C(n,k)$, yield $P(n,k)$. Dividing by $P(k,k)$ yields

$$P(k,k)\, C(n,k) = P(n,k) \tag{4-6}$$

$$C(n,k) = \frac{P(n,k)}{P(k,k)} = \frac{n!}{k!(n-k)!} \tag{4-7}$$

Now that we have characterized some of the ways that we can construct sample spaces, we can now determine how to apply probabilities to the events in the sample space. The first step to achieving this is to assign a set of weights to the events in the sample space. The choice of which weighting factor to apply to what event in the sample space is not a trivial task. One method is to employ observation over a sufficiently long period so that a large sample of all possible outcomes is obtained. This is the so-called "observation, deduction and prediction cycle" and it is useful for developing weights for processes where an underlying model of the process either does not exist or is too complex to yield event weights. This method, sometimes called the "classical probability definition," defines the probability of any event as the following:

$$P(A) = \frac{N_A}{N} \tag{4-8}$$

where $P(A)$ denotes the probability of event A, N_A is the total number of observations where the event A occurred, and N is the total number of observations made. An extension to the classical definition, called the "relative frequency definition," is given as

$$P(A) = \lim_{n \to \infty} \frac{N_A}{N} \approx \frac{N_A}{N} \tag{4-9}$$

The preceding two approaches are often used in practice as a means of establishing a hypothesis about how a process behaves. These methods do indeed define hypotheses because they are both based upon the observation of a finite number of observations. This fact drives the desire to develop axiomatic definitions for the basic laws of probability.

Probability theory, therefore, is based upon a set of three axioms. By definition, the probability of an event is given by a positive number. That is,

$$P(A) \geq 0 \tag{4-10}$$

The following relationship is also defined:

$$P(S) = 1 \tag{4-11}$$

That is, the sum of all of the probabilities of all of the events in the sample space S is equal to 1. This is sometimes called the "certain event." The previous two definitions represent the first two axioms and necessarily restrict the probability of any event to between 0 and 1 inclusively. The third is based upon the property of mutual exclusion, which states that two events are mutually exclusive if, and only if, the occurrence of one of the events positively excludes the occurrence of the other. In set terminology, this states that the intersection of the two events contains no elements; that is, it is the null set. For example, the two events in the coin toss experiment (heads and tails) are mutually exclusive. The third axiom, then, states that the combined probability of events A or B occurring is equal to the sum of their individual probabilities. That is,

$$P(A + B) = P(A) + P(B) \tag{4-12}$$

The reader is cautioned that the expression to the left of the equal sign reads "the probability of event A or event B" whereas the right hand expression reads "the probability of event A plus the probability of event B." This is an important relationship between set theory and the numerical representation of probabilities.

The three axioms of probability are summarized below:

$$\text{I:} \quad P(E) \geq 0 \tag{4-13}$$

$$\text{II:} \quad P(S) = 1 \tag{4-14}$$

$$\text{III:} \quad \text{if } AB \neq 0, \text{ then } P(A + B) = P(A) + P(B) \tag{4-15}$$

In Equation (4-15), the terminology AB is taken as the set A intersected with the set B. The sample space is defined on the total set $\{A_1...A_k\}$ as

$$S = A_1 + A_2 + ... + A_k \tag{4-16}$$

Another important topic in probability theory is that of conditional probability. Consider the following experiment in which we have the events A, B, and AB. The event AB contains the events that are in A and B. Let us say that this event occurs N_{AB} times. Let N_B denote the number of times event B occurs. If we want to know the relative frequency of event A given that event B occurred, we would do the following:

Probability and Statistics   53

$$\text{Relative frequency (A)} = \frac{N_{AB}}{N_B} \tag{4-17}$$

That is, if both events A and B occur (event AB), the number of times event A occurs when event B occurs is found as a fraction of the space of event B where event A intersects (see Figure 4-1). The notation for this relative frequency is denoted P(A/B) and reads as "the conditional probability of event A given event B occurred." If we form the following expression from Equation (4-16),

$$P(A/B) = \frac{N_{AB}/N}{N_B/N} \tag{4-18}$$

and apply Equations (4-9), we obtain the traditional conditional probability definition.

$$P(A/B) = \frac{P(AB)}{P(B)} \tag{4-19}$$

One interesting simplification of Equation (4-18) occurs when the event A is contained in, or is a subset of, event B, such that AB = A. In this case, Equation (4-18) becomes

$$\text{if } AB = A, \text{ then } P(A/B) = \frac{P(A)}{P(B)} \tag{4-20}$$

The independence of two events is defined by the following

$$P(AB) = P(A)\, P(B) \tag{4-21}$$

or

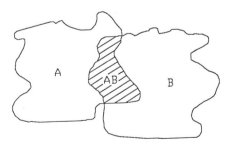

Figure 4-1   Conditional probability space.

$$P(A) = \frac{P(AB)}{P(B)} = P(A/B) \qquad (4\text{-}22)$$

This definition states that the relative number of occurrences of the event A is equal to the relative number of occurrences of event A given event B occurred. In simpler terms, the independence of two or more events indicates that the occurrence of one event does not allow one to infer anything about the occurrence of the other.

Bayes' theorem is stated as follows: If we have a number of mutually exclusive events $B_1, B_2 ... B_N$ whose union define the event space (or more formally, a subset of the sample space), for some experiment, and an arbitrary event A from the sample space, the conditional probability of any event B_k in the set $B_1...B_N$, given that A occurs, is given by

$$P(B_k/A) = \frac{P(B_k)\, P(A/B_k)}{\sum_{i=1}^{N} P(B_i)\, P(A/B_i)} \qquad (4\text{-}23)$$

This result is important because it relates the conditional probability of any event of a subspace of events relative to an arbitrary event of the sample space to the conditional probability of the arbitrary event relative to all of the other events in the subspace. The theorem is a result of the total probability theorem, which states that

$$P(A) = P(A/B_1)\, P(B_1) + ... + P(A/B_k)\, P(B_k) \qquad (4\text{-}24)$$

The theorem holds because the events $B_1...B_k$ are mutually exclusive, and, therefore, $A = AS = A(B_1+B_2...B_k) = AB_1...AB_2$ so that

$$P(A) = P(AB_1) + P(AB_2) + ... + P(AB_k) \qquad (4\text{-}25)$$

Since events $B_1...B_k$ are mutually exclusive, so are events $AB_1...AB_k$. Applying the conditional probability definition of Equation (4-19) to Equation (4-25) yields Equation (4-23). One other relationship that is often useful is the following:

$$P(A + B) = P(A) + P(B) - P(AB) \qquad (4\text{-}26)$$

for any two events A and B. This follows from the discussions of sets of events earlier in this chapter. Since the union of the events A and B yields all sample points in A and B considered together, the sum of the events separately will yield the same quantity plus an extra element for each element in the intersection of the two events. Thus, we must subtract the intersection to form the equality, hence, Equation (4-25). Note that Equation (4-26) is essentially an extended version of Equation (4-15) where AB does not equal the null set.

RANDOM VARIABLES

Thus far, we have been discussing experiments, along with their associated event space, in the context of the probabilities of occurrence of the events. We will now move onto a topic of great importance, which relates the basic probability measures to real world quantities. The concept of a random variable relates the probabilities of the outcomes of an experiment to a range or set of numbers. A random variable, then, is defined as a function whose input values are the events of the sample space and whose outcome is a real number. For example, we could have an experiment in which the outcome is the length of each message that arrives over a communication line. A random variable defined on this experiment could be the number of messages that equaled a certain character count. Often, we want to consider a range of values of the random variables; for instance, the range of messages greater than X_1. This is denoted here as $\{X \leq x_1\}$ where X denotes the random variable and x_1 is a value. We may call this set the event where the random variable X yields a value greater than x_1. Continuing with the previous example, suppose that we had the following outcomes from the message length experiment.

The random variable defined by the number of times the message length of 500 is seen is equal to message 2. The event $\{X > 2000\}$ contains the outcomes of messages 1, 4, 5, and 6.

Random variables may be either discrete of continuous. A discrete random variable is one that is defined on an experiment in which the number of events in the set of outcomes is finite or countably infinite (i.e., it is possible to assign a positive integer to each event, even if there are an infinite number of outcomes). A continuous random variable is one that is defined on an experiment in which the number of possible outcomes is infinite (i.e., defined on the real line). The concept of random variables forms the foundation for the discussion of probability distributions and density functions.

JOINTLY DISTRIBUTED RANDOM VARIABLES

Suppose we have an experiment that has two or more random variables defined on its event space and that we wish to form a random variable that takes into account each of the individual random variables. These are called jointly distributed random variables, and they represent the intersections of the individual random variable event spaces. Jointly distributed random variables are represented with the following notation:

$\{X \leq x_1, Y \leq y_1\}$

Stated simply, joint random variables derive their output from a function whose domain is the set of outcomes for all of the individual random variable domains.

OUTCOME	MESSAGE LENGTH
1	2097
2	500
3	1259
4	5794
5	4258
6	5205

Figure 4-2 Outcomes of the message length experiment.

It should be noted here that more complicated combinations and conditions for the random variable function may be constructed. For example, consider the following random variables

$\{x_1 \leq X \leq x_2\}$ where $x_1 < x_2$
$\{x_1 > X, x_2 < X\}$ where $x_1 > x_2$
$\{x_1 \leq X \leq x_2, y_1 \leq Y \leq y_2\}$ where $x_1 < x_2$ and $y_1 < y_2$

PROBABILITY DISTRIBUTIONS

The concept of a random variable in and of itself does not lend itself to extensive practical use. To remedy this we define a distribution function for each random variable X. The distribution function is typically represented as

$$F(x) = P(X \leq x) \tag{4-27}$$

By its definition, the distribution function assumes values from 0 to 1. Also, the distribution function is nondecreasing as x increases in value. These properties are summarized below.

Property I $\quad \lim_{x \to -\infty} F(x) = 0$
Property II $\quad \lim_{x \to \infty} F(x) = 1$
Property III $\quad F(x_1) \leq F(x_2)$ if $x_1 \leq x_2$

Distribution functions are also called cumulative distribution functions because at any x along the distribution, the area under the curve to the left of x represents the cumulative total of the probabilities of the random variables $\{x \le X\}$. Figure 4-3 shows some example distribution functions.

From the above, it is obvious that these functions are called "distribution functions" because they show exactly how the probability of the random variable is distributed over the range of the random variable.

The distribution function shown in Figure 4-3(a) is a continuous function because it is based upon a continuous random variable. Figure 4-3(b) shows a discrete distribution function that is based upon a discrete random variable.

A joint distribution is one that defines how the probability is associated with each of several random variables. Thus, we can state a function that defines a joint distribution as

$$F(x,y) = P(X \le x, Y \le y) \tag{4-28}$$

This may be interpreted as

$$F(x,y) = P(X \le x \text{ intersected with } Y \le y) \tag{4-29}$$

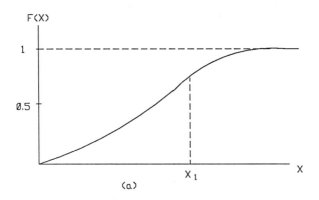

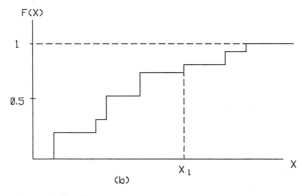

Figure 4-3 Example distribution functions.

We are also interested in the individual distribution functions of X and Y given the joint distribution of Equation (4-28). For instance, the distribution of X given F(x,y), also called the "marginal distribution function" of X corresponding to F(x,y), is given as

$$F_x(x) = \lim_{y \to \infty} F(x,y) \tag{4-30}$$

or

$$F_x(x) = F(x,\infty) \tag{4-31}$$

The same is true for the marginal distribution of Y

$$F_y(y) = F(\infty,y) \tag{4-32}$$

The marginal distributions of the random variables given above result from the definitions of random variables and of distribution functions. Remember that a random variable is defined with a range of values along the real axis and that the distribution function is defined as the cumulative probability that the random variable will attain at least a certain value. The probability then, that the random variable will obtain a value less than infinity is equal to one. Thus the marginal distribution for a random variable given a joint distribution is clear given that:

$$(X \leq x) = (X \leq x, Y \leq \infty) \tag{4-33}$$

DENSITIES

A density function defines the derivative of the distribution function, therefore indicating the rate of change of the probability distribution.

$$f(x) = \frac{d F(x)}{dx} \tag{4-34}$$

The above definition holds for continuous random variables. For discrete random variables, the density function is defined as the discrete probabilities that the random variable equals a specific value for its range of possible values. That is,

$$f(x) = \sum_i P(X \leq x_i) \, \delta(x - x_i) \tag{4-35}$$

where $\delta(x - x_i)$ is a delta function that is 1 when $x = x_i$ and 0 elsewhere.

Probability and Statistics 59

From the above relationships, we can see how the distribution function is formed. For each value of the random variable, we can integrate (for a continuous function) up to that point to find the cumulative probability to that point. The probabilities are summed for discrete functions.

$$F(x) = \int_{-\infty}^{x} f(t)dt \tag{4-36}$$

$$F(x) = \sum_{n=-\infty}^{x} f(n) \tag{4-37}$$

We know from the previous discussions that $F(\infty) = 1$ so that

$$\int_{-\infty}^{\infty} f(t)dt = 1$$

and

$$\sum_{t=-\infty}^{-\infty} f(t) = 1 \tag{4-38}$$

In a manner similar to that shown above for finding the distribution function from the density function for a single random variable, we can find the joint distribution from the joint density. The relationship is given by

$$F(x,y) = \int_{-\infty}^{x} \int_{-\infty}^{y} f(t,u) \, dt du \tag{4-39}$$

Similarly, a discrete distribution can be found from the discrete density

$$F(x,y) = \sum_{i=-\infty}^{x} \sum_{j=-\infty}^{y} f(i,j) \tag{4-40}$$

As with singularly distributed densities, the total area under the probability density function is given by

$$\int_{-\infty}^{\infty} \int_{-\infty}^{\infty} f(x,y) \, dx \, dy = 1 \tag{4-41}$$

Obtaining the density function from the distribution function for a continuous case is given by

$$f(x,y) = \frac{\partial^2 F(x,y)}{\partial x \, \partial y} \qquad (4\text{-}42)$$

We define the marginal density of a jointly distributed random variable as

$$f_Y(y) = \int_{-\infty}^{\infty} f(x,y) \, dx \qquad (4\text{-}43)$$

The independence property is defined on joint distributions as

$$F(x,y) = F_x(x) \, F_y(y) \qquad (4\text{-}44)$$

and for joint densities as

$$f(x,y) = f_X(x) \, f_Y(y) \qquad (4\text{-}45)$$

In some cases, it is necessary to define combined joint distributions in which one of the variables is discrete and the other continuous. The joint density, where y represents the continuous variable and i, the discrete one, is written as

$$f(i,y) = f_{Y/X}(y/i) \, P_X(i) \qquad (4\text{-}46)$$

This expression introduces another important point, that of conditional distributions. For discrete random variables, the conditional function can be defined as the following

$$f_{X/Y}(x/y) = \frac{f(x,y)}{f_Y(y)} \qquad (4\text{-}47)$$

Similarly, we can define the conditional density of y given x. From Equation (4-46). The following results:

$$f(x,y) = f_{X/Y}(x/y) \, f_Y(y) = f_{Y/X}(y/x) f_X(x) \qquad (4\text{-}48)$$

This is a convenient way to relate the conditional densities for the two random variables. If the random variables X and Y are independent, Equation (4-47) becomes

$$f(x,y) = f_X(x) \, f_Y(y) \qquad (4\text{-}49)$$

and the following results

$$f_{Y/X}(y/x) = f_Y(y) \tag{4-50}$$

From Equations (4-43) and (4-48), we can substitute to get

$$f_Y(y) = \int_{-\infty}^{\infty} f_X(x) f_{Y/X}(y/x) \, dx \tag{4-51}$$

and also (for the marginal density of X)

$$f_X(x) = \int_{-\infty}^{\infty} f_Y(y) f_{X/Y}(x/y) \, dy \tag{4-52}$$

Combining now Equations (4-52), (4-51) and (4-48), we obtain Bayes' rule for continuous random variables

$$f_{X/Y}(x/y) = \frac{f_X(x) f_{Y/X}(y/x)}{\int_{-\infty}^{\infty} f_X(x) f_{Y/X}(y/x) \, dx} \tag{4-53}$$

This concludes our discussion on the properties of probability distributions and densities. In the next section, we will explore some methods for obtaining often used statistics about random variables using their distributions and densities.

EXPECTATION

Although both the distribution and density functions of a random variable provide all of the information necessary to describe its behavior, we often wish to have a single quantity (or a small number of them) that provide summary information of the random variable. One such measure is the expected value, or expectation, of a random variable. The expected value is also called the "mean." Expectation for a discrete random variable X is defined as

$$E[x] = \sum_x x \, P(x) \tag{4-54}$$

and for a continuous random variable X with density function $f(x)$ as

$$E[x] = \int_{-\infty}^{\infty} x f(x)\, dx \tag{4-55}$$

Suppose now that we have a function of a random variable X, say $g(X)$. The expectation is given as

$$E[g(x)] = \int_{-\infty}^{\infty} g(x) f(x)\, dx \tag{4-56}$$

for continuous random variables, and as

$$E[g(x)] = \sum_{x} g(x) P(x)$$

for discrete random variables.

If we have jointly distributed random variables, the expectation is defined for discrete random variables as

$$E[g(x,y)] = \sum_{x} \sum_{y} g(x,y) f(x,y) \tag{4-57}$$

and for continuous random variables as

$$E[g(x,y)] = \int_{-\infty}^{\infty}\int_{-\infty}^{\infty} g(x,y) f(x,y)\, dx\, dy \tag{4-58}$$

for the function $g(X,Y)$. The above formulations for expected values are valid if the right-hand sides of the respective equations are less than infinity.

There are a few useful laws relating to expectation that we will now discuss. Suppose that we wish to find the following:

$$E[ax + b] = \int_{-\infty}^{\infty} (ax + b) f(x)\, dx \tag{4-59}$$

The expression on the right becomes

$$a \int_{-\infty}^{\infty} x f(x)\, dx + b \int_{-\infty}^{\infty} f(x)\, dx \tag{4-60}$$

Probability and Statistics   63

From Equations (4-55) and (4-38), Equation (4-60) becomes

$$E[ax + b] = aE[x] + b \tag{4-61}$$

Setting either a or b to zero results in the following

$$E[ax] = aE[x] \tag{4-62}$$

$$E[b] = b \tag{4-63}$$

Now suppose that we have the following

$$E[g(x) + h(x)] = \int_{-\infty}^{\infty} (g(x) + h(x)) f(x) \, dx \tag{4-64}$$

The integral becomes

$$\int_{-\infty}^{\infty} g(x) f(x) \, dx + \int_{-\infty}^{\infty} h(x) f(x) \, dx \tag{4-65}$$

From (4-56), we obtain

$$E[g(x) + h(x)] = E[g(x)] + E[h(x)] \tag{4-66}$$

Similarly for functions of two random variables, we get

$$E[g(x,y) + h(x,y)] = \int_{-\infty}^{\infty}\int_{-\infty}^{\infty} (g(x,y) + h(x,y)) f(x,y) \, dydx \tag{4-67}$$

which becomes

$$\int_{-\infty}^{\infty}\int_{-\infty}^{\infty} g(x,y) f(x,y) \, dydx + \int_{-\infty}^{\infty}\int_{-\infty}^{\infty} h(x,y) f(x,y) \, dydx \tag{4-68}$$

From Equation (4-58), we obtain

$$E[g(x,y) + h(x,y)] = E[g(x,y)] + E[h(x,y)] \tag{4-69}$$

Similarly,

$$E[x+y] = E[x] + E[y] \tag{4-70}$$

Consider the case of two independent random variables X and Y, by Equation (4-58),

$$E[xy] = \int_{-\infty}^{\infty}\int_{-\infty}^{\infty} xy\, f(x,y)\, dydx \tag{4-71}$$

which because of Equation (4-45) becomes

$$\int_{-\infty}^{\infty}\int_{-\infty}^{\infty} xy\, f_X(x)\, f_Y(y)\, dydx \tag{4-72}$$

Separating the integrals by integrands yields

$$\int_{-\infty}^{\infty} x f_X(x)\, dx \int_{-\infty}^{\infty} y f_{Y(y)}\, dy \tag{4-73}$$

From Equations (4-55) and (4-71), we get

$$E[xy] = E[x]\, E[y] \tag{4-74}$$

for the independent random variables X and Y.

For one special function of a random variable, $g(X) = x^n$, the expectation of $g(X)$ is known as the "nth moment" of the random variable X. The first moment of $g(X)$ is defined as the mean of the random variable X for $g(X) = X$. Moments, as defined above, are centered at the origin and are thus called "moments about the origin." A more common and useful definition of moments involve the shifting of the density function so that the mean is centered at the origin. Moments defined as such are called "central moments" because they are defined on density functions that have been centered at the origin. Thus, the function of the random variable becomes

$$g(X) = (x - \mu)^n \tag{4-75}$$

where the mean is given by

$$\mu = E[X] \tag{4-76}$$

The central moment, or moment about the mean, is therefore defined as

$$\mu_n = E[(X-\mu)^n] = \sum_x (x-\mu)^n f(x) \tag{4-77}$$

for the discrete random variable X, and as

$$\mu_n = \int_{-\infty}^{\infty} (x-\mu)^n f(x)\, dx \tag{4-78}$$

for the continuous random variable X.

An important measure of the variability of the distribution of a function about the mean is called the "variance." This measure tells us, loosely speaking, how concentrated the values of the function are relative to the mean. A small variance, therefore, indicates that the probability is that the range of function values are concentrated near the mean, while a large variance suggests that the values are more spread out. The variance of a random variable is defined by its second central moment and represented as

$$\sigma^2 = \text{Var}[X] = \mu_2 = E[X-\mu)^2] = \int_{-\infty}^{\infty} (x-\mu)^2 f(x)\, dx \tag{4-79}$$

Note the usage of several different notations; all are common. For some functions $f(x)$, the integral of Equation (4-79) may be difficult to evaluate. Fortunately, we can derive an alternative expression for the variance as follows:

$$\begin{aligned}
\sigma^2 &= E[(X-\mu)^2] \\
&= E[X^2 - 2X\mu + \mu^2] \\
&= E[X^2] - 2\mu E[X] + \mu^2 \quad \text{by (4.61) + (4.66)} \\
&= E[X^2] - 2\mu^2 + \mu^2 \quad \text{by (4.76)} \\
\sigma^2 &= E[X^2] - \mu^2
\end{aligned} \tag{4-80}$$

The standard deviation of a random variable is defined as the square root of the variance and is denoted as

$$\sigma = \sqrt{\sigma^2} = \sqrt{E[X^2] - \mu^2} \tag{4-81}$$

The covariance of two random variables is a measure of the degree of linear dependence, also called "correlation," of the two variables. The covariance is defined as

$$C_{ov}[X,Y] = E[(X-\mu_x)(Y-\mu_y)] \tag{4-82}$$

If X and Y are independent, the covariance is equal to zero. This results from the following:

$$\begin{aligned}
E[(X-\mu_x)(Y-\mu_y)] &= E[XY - X\mu_y - Y\mu_x + \mu_x\mu_y] \\
&= E[XY] - \mu_y E[X] - \mu_x E[Y] + \mu_x\mu_y \quad \text{(by (4.70))} \\
&= E[XY] - 2\mu_x\mu_y + \mu_x\mu_y \\
&= E[XY] - \mu_x\mu_y \\
C_{ov}[X,Y] &= E[X]E[Y] - \mu_x\mu_y \quad \text{(by (4.74))}
\end{aligned}$$

(4-83)

Equation (4-83) gives a more convenient means for calculating covariance. Two random variables are said to be uncorrelated if $\text{Cov}[X,Y] = 0$.

There are several useful properties of the variance, which we will now discuss; this will be followed by the method for developing a lower bound on the probability for any random variable, given a distance from the mean measured in standard deviations.

From Equations (4-59) and (4-79), we can easily show that

$$\begin{aligned}
\text{VAR}[X+b] &= E[(X+b) - E[X+b])^2] \\
&= E[(X+b) - E[X] - b)^2] \\
\text{VAR}[X+b] &= E[(X-\mu_x)^2] = \text{VAR}[X]
\end{aligned}$$

(4-84)

From Equations (4-62) and (4-79)

$$\begin{aligned}
\text{VAR}[aX] &= E[(aX - E[aX])^2] \\
&= E[a^2(X - E[X])^2] \\
\text{VAR}[aX] &= a^2 E[(X-\mu_x)^2] = a^2 \text{VAR}[X]
\end{aligned}$$

(4-85)

For two jointly distributed random variables X and Y, the variance is defined as

$$\begin{aligned}
\text{VAR}[X+Y] &= E[(X+Y - E[X+Y])^2] \\
&= E[(X-\mu_x + Y-\mu_y)^2] \\
\text{VAR}[X+Y] &= E[(X-\mu_x)^2] + E[(Y-\mu_y)^2] + 2E[(X-\mu_x)(Y-\mu_y)] \\
&= \text{VAR}[X] + \text{VAR}[Y] + 2\text{Cov}[X,Y]
\end{aligned}$$

(4-86)

Given any random variable, it is possible to derive an expression that defines the minimum probability of a random variable lying within k standard deviations of its mean. The theorem is known as Chebyshev's Theorem, and is stated as follows:

Probability and Statistics   67

$$P((\mu - k\sigma) < X < (\mu + k\sigma)) \geq 1 - \frac{1}{k^2} \tag{4-87}$$

Equation (4-87) can be derived as follows. From Equation (4-79),

$$\sigma^2 = \int_{-\infty}^{\infty} (x - \mu)^2 f(x) dx$$

$$\sigma^2 = \int_{-\infty}^{\mu - k\sigma} (x - \mu)^2 f(x) dx + \int_{\mu - k\sigma}^{\mu + k\sigma} (x - \mu)^2 f(x) dx + \int_{\mu + k\sigma}^{\infty} (x - \mu)^2 f(x) dx$$

Because the middle integral is positive or zero, we can remove it from the expression to get

$$\sigma^2 \geq \int_{-\infty}^{\mu - k\sigma} (x - \mu)^2 f(x) dx + \int_{\mu + k\sigma}^{\infty} (x - \mu)^2 f(x) dx \tag{4-88}$$

Within the ranges

$$x \geq \mu + k\sigma \quad \text{and} \quad x \leq \mu - k\sigma$$

we have

$$|x - \mu| \geq k\sigma$$

so that

$$(x - \mu)^2 \geq (k\sigma)^2$$

Thus we can substitute into Equation (4-88)

$$\sigma^2 \geq \int_{-\infty}^{\mu - k\sigma} (k\sigma)^2 f(x) dx + \int_{\mu + k\sigma}^{\infty} (k\sigma)^2 f(x) dx$$

and divide to get

$$\frac{\sigma^2}{(k\sigma)^2} \geq \int_{-\infty}^{\mu - k\sigma} f(x) dx + \int_{\mu + k\sigma}^{\infty} f(x) dx \tag{4-89}$$

From Equation (4-36) we have

$$P((\mu - k\sigma) < X < (\mu + k\sigma)) = \int_{\mu-k\sigma}^{\mu+k\sigma} f(x)dx$$

and from Equations (4-89) and (4-38), we have

$$\int_{-\infty}^{\mu-k\sigma} f(x)dx + \int_{\mu+k\sigma}^{\infty} f(x)dx = \int_{-\infty}^{\infty} f(x)dx - \int_{\mu-k\sigma}^{\mu+k\sigma} \leq \frac{1}{k^2}$$

$$= 1 - \int_{\mu-k\sigma}^{\mu+k\sigma} f(x)dx \leq \frac{1}{k^2}$$

Thus, Equation (4-87) results.

$$P((\mu - k\sigma) < X < (\mu + k\sigma)) = \int_{\mu-k\sigma}^{\mu+k\sigma} f(x)dx \geq 1 - \frac{1}{k^2}$$

SOME EXAMPLE PROBABILITY DISTRIBUTIONS

In this section, we will examine some discrete and some continuous probability distributions that will help to solidify the basic probability theory of the previous sections. Many of these distributions are commonly used to model real world processes and to help arrive at estimates for quantities of interest in real world systems. We will discuss the properties of each distribution and we will also discuss the process of deriving random deviates, given a certain distribution that models a real world process.

The simplest of all probability distributions is the discrete uniform distribution. Such a distribution states that all values of the random variable are equally probable and depend only upon the number of possible outcomes of the experiment. The density function for the uniform distribution is given as

$$f(x) = \frac{1}{k} \qquad x = x_1, x_2, \ldots x_k \tag{4-90}$$

where k is the number of possible outcomes. The experiment where the random variable $X = P(n)$, $n = 1$ to 6, and the event is the toss of a die that results in a discrete uniform probability distribution. A plot of the density function for the uniform distribution is shown in Figure 4-4.

Probability and Statistics   69

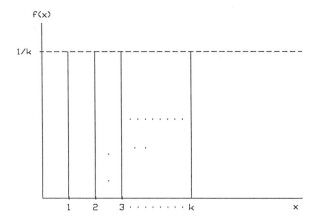

Figure 4-4   Uniform density function.

The mean of the uniform distribution is found by Equation (4-54) and is given by

$$E[X] = \sum_{i=1}^{k} x_i \left(\frac{1}{k}\right) \tag{4-91}$$

The standard deviation is found by

$$\sigma^2 = E[(X-\mu)^2] = \sum_{i=1}^{k} (x_i - \mu)^2 f(x_i)$$

$$\sigma^2 = \sum_{i=1}^{k} \frac{(x_i - \mu)^2}{k} \tag{4-92}$$

The concept of a Bernoulli trial is important in many discrete distributions. A Bernoulli trial is an experiment in which the outcome can be only success or failure. Random variables defined on successive Bernoulli trials make up several of the discrete density functions to be discussed.

An important discrete probability distribution is the binomial distribution. This distribution results from experiments in which there are only two possible outcomes of an experiment, such as a coin toss. For the distribution, one outcome is chosen to represent success and the other failure. A binomial experiment also requires that the probability of success remain constant for successive trials, that the trials themselves are independent, and that each experimental outcome results in success or failure. Since the trials are independent, the total probability for an experiment with x successes and n trials can be found by simply multiplying the probability of each event [see Equation (4-21)]. If the probability of success is given as p and if $q = 1 - p$, we have

$$P(x \text{ successes in n trials}) = p^x q^{(n-x)} \tag{4-93}$$

For the binomial distribution, we want to find the number of successes in n independent trials, given that we know the probability of success for any individual trial. The number of successes in n trials is a combination as given in Equation (4-5). The probability of x successes in n trials, then, is the expression for the binomial distribution:

$$f(x) = C(n,x) \, p^x q^{(n-x)} \quad x = 1, 2, \ldots n \tag{4-94}$$

Suppose we have a binomial experiment in which the outcomes of n experiments can be used to represent the random variable X, which denotes the number of successes in n trials. Thus, by Equation (4-70) and by the definition of binomial random variables,

$$E[X] = E[X_1] + E[X_2] + \ldots + E[X_n]$$
$$E[X] = np \tag{4-95}$$

Since the variance of any of the individual experiments is pq, by Equation (4-86), the variance of a binomial density can be found to be

$$VAR[X] = VAR[X_1] + VAR[X_2] + \ldots VAR[X_n]$$
$$VAR[X] = npq$$

Suppose that we wish to know how many Bernoulli trials occur before the first success in a sequence of trials occurs. If the first trial yields a success and the probability of success for any trial is p, the probability of the random variable X is p. If the probability of failure is given as $q = 1 - p$ and we have success on the second trial, we obtain a probability of pq. Extending to $k - 1$ failures before an eventual success, we obtain what is know as the "geometric distribution," where

$$f(k) = pq^{k-1} \quad k = 1, 2, \ldots \tag{4-96}$$

Finding the expected value of the geometric density function is a bit tricky but can be accomplished as follows. By (4-54),

$$E[X] = \sum_{i=1}^{\infty} i\, pq^{(i-1)}$$

$$E[X] = p \sum_{i=1}^{\infty} i\, q^{(i-1)}$$

$$E[X] = p \sum_{i=0}^{\infty} \frac{d}{dq} q^i$$

$$E[X] = p \frac{d}{dq} \sum_{i=0}^{\infty} q^i$$

$$E[X] = p \frac{d}{dq} \left[\frac{1}{1-q} \right]$$

$$E[X] = \frac{p}{(1-q)^2}$$

$$E[X] = \frac{1}{p} \tag{4-97}$$

The fifth line of the derivation above results because the value of q is less than or equal to 1, thus the summation converges to $1/(1-q)$. The variance of the geometric density is not derived here but is given as

$$\text{VAR}[X] = \frac{q}{p^2} \tag{4-98}$$

See [Trivedi 1982] for a derivation of this quantity.

A widely used discrete density function that is useful for deriving statistics about the number of successes during a given time period is the "Poisson" distribution. The Poisson density is popular mainly because it describes many real word processes very well. In computer systems, requests for jobs at a CPU are often represented by a Poisson process. The Poisson density function is defined as

$$f(x) = \frac{e^{-\mu} \mu^x}{x!} \qquad x = 0, 1, \ldots \tag{4-99}$$

where the parameter μ is defined as the average number of successes during the interval. Several conditions must prevail for a Poisson density function to exist. These are that the successes for one interval are independent of the successes in any other interval, that the probability of a success during an interval of extremely short length is near zero, and that the probability of only one success during a short interval depends only upon the length of the interval. Interestingly, the mean of the Poisson distribution itself is part of its definition. The expected value can be found as:

72 Modeling and Analysis of Local Area Networks

$$E[X] = \sum_{x=0}^{\infty} x \frac{e^{-\mu}\mu^x}{x!}$$

$$E[X] = \sum_{x=1}^{\infty} \frac{xe^{-\mu}\mu^x}{x!}$$

$$E[X] = \mu \sum_{x=1}^{\infty} \frac{e^{-\mu}\mu^{x-1}}{(x-1)!}$$

Now, if we let $y = x - 1$, we arrive at a summation of the density function from 1 to infinity, which by Equation (4-38) is equal to 1:

$$E[X] = \mu \sum_{y=0}^{\infty} \frac{e^{-\mu}\mu^y}{y!}$$

$$E[X] = \mu \tag{4-100}$$

The variance of the Poisson distribution can be found by first finding $E[X(X-1)]$ and then using the result in Equation (4-80).

$$E[X(X-1)] = \sum_{x=0}^{\infty} \frac{x(x-1)e^{-\mu}\mu^x}{x!}$$

The first two terms of this summation are zero, so we have

$$E[X(X-1)] = \sum_{x=2}^{\infty} \frac{x(x-1)e^{-\mu}\mu^x}{x!}$$

$$E[X(X-1)] = \sum_{x=2}^{\infty} \frac{e^{-\mu}\mu^{x-2}\mu^2}{(x-2)!} \tag{4-101}$$

By Equation (4-38), and, if we let $x = y + 2$, we get

$$E[X(X-1)] = \mu^2 \sum_{y=0}^{\infty} \frac{e^{-\mu}\mu^y}{y!} = \mu^2$$

By Equation (4-80), we get

$$\sigma^2 = E[X^2] - \mu^2$$
$$\sigma^2 = E[X^2] - E[X] + E[X] - \mu^2$$

Probability and Statistics 73

By Equation (4-70), we get

$$\sigma^2 = E[X^2 - X] + E[X] - \mu^2$$
$$\sigma^2 = E[X(X-1)] + E[X] - \mu^2$$

By Equation (4-101), we get

$$\sigma^2 = \mu^2 + \mu - \mu^2 = \mu \qquad (4\text{-}102)$$

The previous density functions provide some examples of the more common discrete random variables. Distributions such as this are useful for modeling real world processes in which the quantities of interest are countable items.

In addition to the basic discrete density functions described earlier, there are several continuous densities that are often used. Continuous density functions are characterized by the fact that the value of $f(x)$ at any point x is zero. However, the probability that any value x lies between x and some small delta is approximately $f(x)$ times the delta value.

One of the most important continuous probability distributions, and probably the most widely used, is the "normal," or "Gaussian" distribution.

The density function of a normal random variable X is given as

$$f(x) = \frac{1}{\sqrt{2\pi}\sigma} e^{-(x-\mu)^2/2\sigma^2} \qquad (4\text{-}103)$$

Figure 4-5 shows a few normal curves (also known as bell curves because of their bell-like shapes). The flatter curve has a larger standard deviation than the thinner curves. The expected value of the normal curve is found as follows. By Equation (4-55),

$$E[X] = \int_{-\infty}^{\infty} \frac{x}{\sqrt{2\pi}\sigma} e^{-(x-\mu)^2/2\sigma^2} dx$$

If we substitute the following

$$y = \frac{x-\mu}{\sigma}$$

$$dx = \sigma dy$$

we get

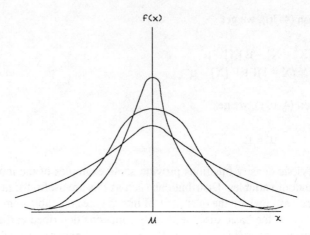

Figure 4-5 A few normal curves.

$$E[X] = \frac{1}{\sqrt{2\pi}} \int_{-\infty}^{\infty} (y\sigma + \mu) e^{-y^2/2} dy$$

$$E[X] = \frac{\sigma}{\sqrt{2\pi}} \int_{-\infty}^{\infty} y\, e^{-y^2/2} dy + \frac{\mu}{\sqrt{2\pi}} \int_{-\infty}^{\infty} e^{-y^2/2} dy$$

If, in the second integral, we replace y by the following:

$$y = (x - \mu)/\sigma$$

$$dy = \frac{1}{\sigma} dx$$

we obtain the expression for the integral of a density function that is equal to 1:

$$E[X] = \frac{\sigma}{\sqrt{2\pi}} \int_{-\infty}^{\infty} (y e^{-y^2/2} dy + \frac{\mu}{\sqrt{2\pi}\sigma} \int_{-\infty}^{\infty} e^{-(x-\mu)^2/2\sigma^2} dx$$

$$E[X] = \frac{\sigma}{\sqrt{2\pi}} \int_{-\infty}^{\infty} y\, e^{-y^2/2} dy + \mu$$

The expression in the remaining integral is an odd function because of the presence of y. Since an odd function integrated over symmetric limits is zero, the mean becomes

$$E[X] = \mu \tag{4-104}$$

We can find the variance of the normal distribution as follows:

$$E[(X-\mu)^2] = \int_{-\infty}^{\infty} \frac{1}{\sqrt{2\pi}\,\sigma} (x-\mu)^2 e^{-(x-\mu)^2/2\sigma^2} dx$$

Making the same substitution as before, we get

$$E[(X-\mu)^2] = \frac{\sigma^2}{\sqrt{2\pi}} \int_{-\infty}^{\infty} y^2 e^{-y^2/2} dy$$

Now, we can integrate by parts:

Let $\quad u = y \quad, \quad v = -e^{-y^2/2}$

and $\quad du = dy \quad, \quad dv = ye^{-y^2/2}$

then $\quad E[(X-\mu)^2] = \frac{\sigma^2}{\sqrt{2\pi}} \left[-ye^{-y^2/2} \Big|_{y=-\infty}^{\infty} + \int_{-\infty}^{\infty} e^{-y^2/2} dy \right]$

The second integral can be shown to be

$$\int_{-\infty}^{\infty} e^{-y^2/2} dy = \sqrt{2\pi}$$

so we have

$$E[(X-\mu)^2] = \frac{\sigma^2}{\sqrt{2\pi}} [0 + \sqrt{2\pi}]$$
$$E[(X-\mu)^2] = \sigma^2 \qquad (4\text{-}105)$$

For the normal curve, the mean occurs at the mode, which is defined as the value that appears most in the distribution.

Finding the probability that a normally distributed random variable falls between two values requires the solution of the integral:

$$P(x_1 < X < x_2) = \frac{1}{\sqrt{2\pi}\,\sigma} \int_{x_1}^{x_2} e^{-(x-\mu)^2/2\sigma^2} dx \qquad (4\text{-}106)$$

This integral is not easily solvable and is best approached using numerical means. In order to be useful, however, we would need to generate a table for each value of mean and standard deviation. We would like to avoid this by having only one standard normal curve. If we make the substitution

$$z = \frac{x - \mu}{\sigma}, \quad dx = \sigma dz$$

in the previous equation, we obtain

$$\frac{1}{\sqrt{2\pi}\,\sigma} \int_{z_1}^{z_2} e^{-z^2/2} dz \qquad (4\text{-}107)$$

where

$$z_1 = \frac{x_1 - \mu}{\sigma} \quad \text{and} \quad z_2 = \frac{x_2 - \mu}{\sigma}$$

The expression is equivalent to a normal distribution of mean equal to 0 and standard deviation of 1. Thus, we can transform any normal distribution into the standard normal curve with 0 mean and a standard deviation of 1. For example, let's compute the probability that any normally distributed random variable falls within one standard deviation of the mean. To do so, we need to generate some sort of table for the standard normal distribution. Table 4-1 gives values for the standard normal distribution integrated from minus infinity to x.

$$\frac{1}{\sqrt{2\pi}\,\sigma} \int_{-\infty}^{x} e^{-x^2/2} dx$$

Let $\quad x_1 = \mu - \sigma \quad$ and $\quad x_2 = \mu + \sigma$

then $\quad z_1 = \dfrac{\mu - \sigma - \mu}{\sigma} \quad$ and $\quad z_2 = \dfrac{\mu + \sigma - \mu}{\sigma}$

so $\quad z_1 = -1 \quad$ and $\quad z_2 = 1$

the left hand side of Equation (4-106) can be rewritten as

$$P(x_1 < X < x_2) = P(X < x_2) - P(X < x_1)$$

The values from the table for $z = -1$ and 1, respectively, are

Table 4-1 Standard Normal Curve Values

x	x+0.00	x+0.01	x+0.02	x+0.03	x+0.04	x+0.05	x+0.06	x+0.07	x+0.08	x+0.09
0.0	0.50000	0.50401	0.50800	0.51199	0.51597	0.51996	0.52394	0.52794	0.53194	0.53594
0.1	0.53993	0.54391	0.54790	0.55187	0.55585	0.55981	0.56378	0.56773	0.57168	0.57562
0.2	0.57955	0.58348	0.58740	0.59130	0.59520	0.59909	0.60297	0.60684	0.61070	0.61455
0.3	0.61839	0.62221	0.62603	0.62983	0.63362	0.63739	0.64116	0.64490	0.64864	0.65236
0.4	0.65607	0.65976	0.66343	0.66709	0.67074	0.67436	0.67798	0.68157	0.68515	0.68871
0.5	0.69225	0.69578	0.69929	0.70277	0.70624	0.70970	0.71313	0.71654	0.71993	0.72331
0.6	0.72666	0.72999	0.73331	0.73660	0.73987	0.74312	0.74635	0.74956	0.75274	0.75591
0.7	0.75905	0.76217	0.76527	0.76834	0.77139	0.77442	0.77743	0.78041	0.78337	0.78631
0.8	0.78923	0.79212	0.79498	0.79783	0.80065	0.80344	0.80622	0.80897	0.81169	0.81439
0.9	0.81707	0.81972	0.82234	0.82495	0.82753	0.83008	0.83261	0.83512	0.83760	0.84006
1.0	0.84248	0.84487	0.84724	0.84959	0.85192	0.85421	0.85649	0.85874	0.86097	0.86317
1.1	0.86535	0.86751	0.86964	0.87174	0.87383	0.87589	0.87793	0.87994	0.88193	0.88389
1.2	0.88584	0.88775	0.88965	0.89152	0.89337	0.89520	0.89701	0.89879	0.90055	0.90228
1.3	0.90400	0.90569	0.90736	0.90901	0.91063	0.91224	0.91382	0.91538	0.91692	0.91844
1.4	0.91994	0.92142	0.92287	0.92431	0.92572	0.92712	0.92849	0.92985	0.93118	0.93250
1.5	0.93379	0.93507	0.93633	0.93756	0.93878	0.93998	0.94117	0.94233	0.94348	0.94460
1.6	0.94571	0.94681	0.94788	0.94894	0.94998	0.95100	0.95201	0.95300	0.95397	0.95493
1.7	0.95587	0.95679	0.95770	0.95860	0.95947	0.96034	0.96119	0.96202	0.96284	0.96364
1.8	0.96443	0.96521	0.96597	0.96672	0.96745	0.96817	0.96888	0.96957	0.97026	0.97093
1.9	0.97158	0.97223	0.97286	0.97348	0.97409	0.97468	0.97527	0.97584	0.97640	0.97695
2.0	0.97749	0.97803	0.97855	0.97906	0.97956	0.98005	0.98053	0.98100	0.98146	0.98191
2.1	0.98235	0.98279	0.98321	0.98363	0.98403	0.98443	0.98482	0.98520	0.98557	0.98593
2.2	0.98629	0.98664	0.98698	0.98731	0.98764	0.98795	0.98827	0.98857	0.98887	0.98916
2.3	0.98944	0.98972	0.98999	0.99025	0.99051	0.99077	0.99101	0.99125	0.99149	0.99172
2.4	0.99194	0.99216	0.99237	0.99258	0.99279	0.99298	0.99318	0.99337	0.99355	0.99373
2.5	0.99391	0.99408	0.99424	0.99441	0.99456	0.99472	0.99487	0.99502	0.99516	0.99530
2.6	0.99543	0.99556	0.99569	0.99582	0.99594	0.99606	0.99618	0.99629	0.99640	0.99650
2.7	0.99661	0.99671	0.99681	0.99690	0.99700	0.99709	0.99717	0.99726	0.99734	0.99742
2.8	0.99750	0.99758	0.99765	0.99773	0.99780	0.99787	0.99793	0.99800	0.99806	0.99812
2.9	0.99818	0.99824	0.99829	0.99835	0.99840	0.99845	0.99850	0.99855	0.99859	0.99864
3.0	0.99868	0.99873	0.99877	0.99881	0.99885	0.99888	0.99892	0.99896	0.99899	0.99902
3.1	0.99906	0.99909	0.99912	0.99915	0.99918	0.99920	0.99923	0.99926	0.99928	0.99930
3.2	0.99933	0.99935	0.99937	0.99939	0.99942	0.99944	0.99945	0.99947	0.99949	0.99951
3.3	0.99953	0.99954	0.99956	0.99957	0.99959	0.99960	0.99962	0.99963	0.99964	0.99966
3.4	0.99967	0.99968	0.99969	0.99970	0.99971	0.99972	0.99973	0.99974	0.99975	0.99976
3.5	0.99977	0.99978	0.99978	0.99979	0.99980	0.99981	0.99981	0.99982	0.99983	0.99983
3.6	0.99984	0.99984	0.99985	0.99986	0.99986	0.99987	0.99987	0.99988	0.99988	0.99988
3.7	0.99989	0.99989	0.99990	0.99990	0.99990	0.99991	0.99991	0.99991	0.99992	0.99992
3.8	0.99992	0.99992	0.99993	0.99993	0.99993	0.99993	0.99994	0.99994	0.99994	0.99994
3.9	0.99994	0.99995	0.99995	0.99995	0.99995	0.99995	0.99995	0.99996	0.99996	0.99996
4.0	0.99996	0.99996	0.99996	0.99996	0.99997	0.99997	0.99997	0.99997	0.99997	0.99997

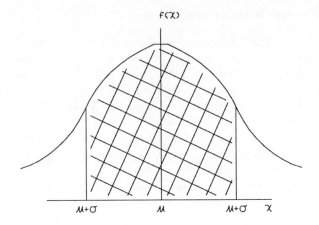

Figure 4-6 The probability of a normal event within the variance.

P (z < −1) = 0.1587

P (z < 1) = 0.8413

Therefore, we have

P ((μ − σ) < x < (μ + σ)) = P (z< 1) − P (z< − 1) = 0.6826

Figure 4-6 shows the area under the normal curve in which we are interested.

A simpler continuous distribution, the exponential, is important in queuing theory and therefore is discussed here. Its main attraction is that it has the Markovian property, which states that the probability of occurrence of an event is completely independent of the history of the experiment. This characteristic is also called the "memoryless" property. The expression for an exponential distribution is given as

$$f(x) = \begin{cases} \lambda e^{-\lambda x} & x > 0 \\ 0 & \text{otherwise} \end{cases} \qquad (4\text{-}108)$$

The graph of an exponential curve is shown in Figure 4-7.

We will see later that the exponential distribution, because of its Markovian property, will be useful for representing service time distributions in queuing systems.

Suppose that the time a computer user spends at a system terminal is exponentially distributed over time. The probability that the user will be at a terminal for n minutes is given as

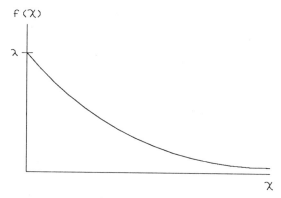

Figure 4-7 Exponential distribution.

$$P(X \geq n) = \int_n^\infty f(x)dx$$

$$P(X \geq n) = \int_n^\infty \lambda e^{-\lambda x} dx$$

$$P(X \geq n) = e^{-n\lambda} \tag{4-109}$$

The probability distribution function for the exponential function is given as

$$F(x) = \begin{cases} 1 - e^{-\lambda x} & x > 0 \\ 0 & \text{otherwise} \end{cases} \tag{4-110}$$

As with any distribution function, we can find the result found earlier by picking the point at n, representing the probability that the user will be at a terminal for less than n minutes, and use Equation (4-14) to find the probability of the complementary event (see also Figure 4-8):

$$P(X \geq n) = 1 - F(n)$$

$$P(X \geq n) = 1 - (1 - e^{-nx})$$

$$P(X \geq n) = e^{-nx}$$

The mean of an exponential random variable is found as

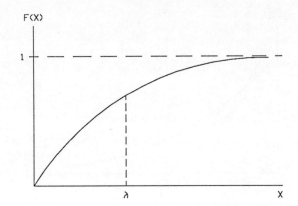

Figure 4-8 Cumulative distribution function for an exponential random variable.

$$E[x] = \int_0^\infty x e^{-\lambda x}\, dx$$

If we let

$$u = x \quad \text{and} \quad v = -e^{-\lambda x}$$

$$du = dx \quad \text{and} \quad dv = \lambda e^{-\lambda x}\, dx$$

and integrate by parts where

$$\int_a^b u\, dv = uv \Big|_a^b - \int_a^b v\, du$$

we obtain

$$E[X] = -x e^{-\lambda x} \Big|_0^\infty - \int_0^\infty -e^{-\lambda x}\, dx$$

Because

$$\lim_{x \to \infty} x e^{-\lambda x} = 0$$

$$E[X] = 0 - \int_0^\infty -e^{-\lambda x} dx$$

$$E[X] = \frac{1}{\lambda} e^{-\lambda x} \Big|_0^\infty$$

$$E[X] = \lim_{x \to \infty} \frac{1}{\lambda} e^{-\lambda x} + \frac{1}{\lambda}$$

$$E[X] = \frac{1}{\lambda} \tag{4-111}$$

We may find the variance of the exponential random variable as follows

$$VAR[X] = E[(X - \mu)^2] = \int_{-\infty}^{\infty} (x - \mu)^2 \lambda e^{-\lambda x} dx$$

$$VAR[X] = \int_0^\infty \lambda x^2 e^{-\lambda x} dx - 2 \int_0^\infty x e^{-\lambda x} dx + \frac{1}{\lambda} \int_0^\infty e^{-\lambda x} dx \tag{4-112}$$

To solve this last equation, we introduce the gamma function, denoted as

$$\text{Gamma}[\propto] = \int_0^\infty x^{\propto - 1} e^{-x} dx$$

The gamma function can be solved for a positive value of the parameter to yield

$$\text{Gamma}[n] = (n - 1)! \tag{4-113}$$

We can now use Equation (4-113) to help find the solution to Equation (4-112) and arrive at the variance for an exponential distribution

$$VAR[X] = \frac{2}{\lambda^2} - \frac{2}{\lambda^2} + \frac{1}{\lambda^2}$$

$$VAR[X] = \frac{1}{\lambda^2} \tag{4-114}$$

82 Modeling and Analysis of Local Area Networks

The exponential density function is often used to represent the service time of a server at the end of a waiting line. In some cases, it is desirable to represent several identical servers with a single density function whose statistics are the same as for a single equivalent exponential server. The distribution that satisfies these conditions is called the "Erlang distribution" and is given as

$$f(x) = \begin{cases} \dfrac{\lambda k \, (\lambda k \, x)^{k-1} \, e^{-\lambda k x}}{(k-1)!} & x > 0 \\ 0 & \text{otherwise} \end{cases} \tag{4-115}$$

with parameters λ and k. Figure 4-9 shows a graph of the Erlang density function for various values of k for a given value of λ. The mean and standard deviation of the Erlang density function are given as

$$E[X] = \frac{1}{\lambda} \tag{4-116}$$

$$\text{VAR}[X] = \frac{1}{k \lambda^2} \tag{4-117}$$

The probability distribution function is given as

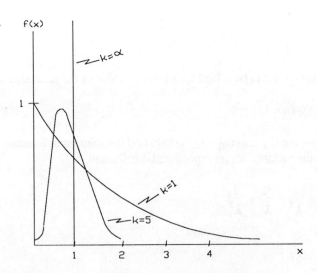

Figure 4-9 Erlang density function for $\lambda = 1$ and $k = 1, 5$ and ∞.

$$F(x) = 1 - e^{-\lambda kx} \left[1 + \sum_{i=0}^{k-1} \frac{(k\lambda)^i}{i!} \right]$$

(4-118)

It is important to note that the expected value for the Erlang density is the same as for an exponential with the same parameter lambda and is independent of the number of Erlangian servers (e.g., parallel servers).

SUMMARY

This chapter has introduced some of the basic probability concepts that are useful for understanding and analyzing queuing network models. Many additional, more complex probability densities are known to be useful for representing certain types of real world processes. These are beyond the scope of this text but may be found in many probability and statistics texts, including [Trivedi 1982], [Allen 1978], [Papoulis 1984], [Walpole 1978] and [Drake 1967]. The densities presented herein, however, are commonly used in queuing analysis due to their applicability to many arrival and service processes and because of their relative computational simplicity.

REFERENCES

Allen, A. D. *Probability, Statistics and Queuing Theory with Computer Science Applications*, Academic Press Inc., Orlando, FL, 1978.

Drake, A. W. *Fundamentals of Applied Probability Theory*, McGraw-Hill Book Company, New York, NY, 1967.

Papoulis, A. *Probability, Random Variables, and Stochastic Processes*, McGraw-Hill Book Company, New York, NY, 1984.

Trivedi, K. *Probability and Statistics with Reliability, Queueing and Computer Science Applications*, Prentice Hall, Englewood, NJ, 1982.

Walpole, R. E. and R. H. Myers. *Probability and Statistics for Engineers and Scientists*, Macmillan Publishing Company, New York, NY, 1978.

5

Simulation Analysis

Simulation is the realization of a model for a system in computer executable form. That is, the model of the real world system has been translated into a computer simulation language. The computer realization provides a vehicle to conduct experiments with the model in order to gain insight into the behavior and makeup of the system or to evaluate alternatives. Simulations, to be effective, require a precise formulation of the system to be studied, correct translation of this formulation into a computer program, and interpretation of the results.

Simulation is usually undertaken because the complexity of systems of interest defy use of simpler mathematical means. This complexity may arise from inherent stochastic processes in the system, complex interactions of elements that lack mathematical formulations, or the sheer interactability of mathematical relationships that result from the system's equations and constraints. Because of these constraints and other reasons such as modeler preference, simulation is often the tool for evaluation. Simulation provides many potential benefits to the modeler: it makes it possible to experiment and study the myriad complex internal interactions of a particular system, with the complexity and therefore benefit level left up to the modeler.

Simulation allows for the sensitivity analysis of the system by providing means to alter the model and observe the effects it has on the system's behavior. Through simulation we can often gain a better understanding of the real system. This is because of the detail of the model and the modeler's need to independently understand the system in order to faithfully construct a simulation of it. The process of learning about the system in order to simulate it many times will lead to suggestions for change and improvements. The simulation then provides a means to test these hypotheses. Simulation often leads to a better understanding of the importance of various elements of a system and how they interact with each other. It provides a laboratory environment in which we can study and analyze many alternatives and their impact well before a real system even exists or, if one exists, without disturbing or perturbing it. Simulation enables the modeler to study dynamic systems in real, compressed, or expanded time, providing a means to examine details of situations and processes that otherwise could not be performed.

Finally, it provides a means to study the effects on an existing system of adding new components, services, etc., without testing them in the system. This provides a means to discover bottlenecks and other problems before we actually expend time and capital to perform the changes.

Simulation has been used for a wide variety of purposes as can be seen from the diversity of topics covered at annual simulation symposiums. Simulation easily lends itself to many fields such as business, economics, marketing, education, politics, social sciences, behavioral sciences, natural sciences, international relations, transportation, war gaming, law enforcement, urban studies, global systems, space systems, computer design and operations, and myriad others.

Up to this point we have used "system" to describe the intended modeled entity. In the context of simulation it is used to designate a collection of objects with a well-defined set of interactions between them. A bank teller interacts with the line of customers, and the job the teller does may be considered a system in this context, with the customers and tellers forming the object collection and the functions performed by each (deposit, withdrawal) as the set of interactions.

Systems in nature are typically described as being continuous or discrete, where these terms imply the behavior of the variables associated with describing the system. They provide us, the modelers, with a context in which to place the model and to build on. In both cases the typical relation of variables is built around time. In the case of the discrete model, time is assumed to step forward in fixed intervals determined by the events of occurrence versus some formulation, and in the continuous model, the variables change continually as time ticks forward. For example, with the bank, if the variable of interest is the number of customers waiting for service, we have a dependent discrete "counting" sequence. On the other hand, if we are looking at a drive-up bank teller and are interested in the remaining fuel in each vehicle and the average, we could model the gasoline consumption as a continuous variable dependent on the time in line until exiting.

Systems can possess both discrete and continuous variables and still be modeled. In reality, this is frequently the case. Another consideration in defining a system is the nature of its processes. Processes, whether they are discrete or continuous, can have another feature, that of being deterministic or stochastic. A deterministic system is where, given an input x and initial conditions I, you will always derive the same output $y=f(x,I)$. That is, if we were to perform the same process an infinite number of times, with the same inputs and same initial state of the process, we would always realize the same result. On the other hand, if the system were stochastic, this would not hold. For the same system with input held at X and initial state held at I, we could have the output Y take on one of many possible outputs. This is based on the "random" nature of stochastic processes. That is, they will be randomly distributed over the possible outcomes. For example, if the bank teller system is described as a discrete system, we are assuming that the service time of the server is exactly the same and the arrival rate of customers is fixed and nonvarying. On the other hand, if the same system is given some reality, we all know that service is random based on the job the tellers must perform and how they perform it. Likewise customers do not arrive in perfect order; they arrive randomly as they please. In both cases the model will give vastly different results.

Simulation Process

The use of a digital computer to perform modeling and run experiments has been a popular technique for quite some time. In this environment simulation can make systematic studies of problems that cannot be studied by other techniques. The simulation model describes the system in terms of the elements of interests and their interrelationships. Once completed, it provides a laboratory in which to carry out many experiments on these elements and interactions.

Simulation programs, as with generic modeling, require discrete phases to be performed in order to realize its full potential. They are:

1. Determine that the problem requires simulation.
2. Formulate a model to solve the problem.
3. Formulate a simulation model of the problem.
4. Implement the model in a suitable language.
5. Design simulation experiments.
6. Validate the model.
7. Perform experiments.

The typical simulation model project will spend most of its time in phases 2, 3, and 4 because of the complexities associated with formulating the model and the conversion to simulation format and implementation in a language. Model formulation deals with the definition of critical elements of the real world system and their interactions. Once these critical elements have been identified and defined (mathematically, behaviorally, functionally) and their interactions (cause and effect, predecessor and successor, dependencies and independencies, data flows, and control flow) are defined in terms of their essence, simulation model development flows into and along with systems model definition. That is, as we develop a system model we can often directly define the simulation model structure.

An important aspect of this model development is the selection of a proper level of simulation, which is directly proportional to the intended purpose of the performance evaluation, the degree of understanding of the system, its environment, and the output statistics required. On one extreme, for example, we could model our bank teller system down to the level of modeling all his or her actions. Or, on the other hand, we could model the teller service as strictly a gross estimate of time to perform service regardless of the type of service. The level to choose would be dependent on what is to be examined. In the first example, we may wish to isolate the most time-consuming aspect(s) of their functions so that we could develop ways to improve them. At the second level possibly all we wish to determine is based on the customer load, the average teller service time, and the optimal number of tellers to have on duty and when.

The intent of the performance measure drives us directly to a simulation level of detail, which typically falls somewhere in between the two extremes: too low or too high to be useful. In most cases, however, we as modelers do not or cannot always foresee how the level of detail of all components can influence the model's ultimate usefulness. A solution typically used to cope with such uncertainties is to construct the model in a modular fashion, allowing each component to migrate to the level consistent with its intent and overall impact on the simulation and system. What this typically drives us to is top-down model development with each layer being refined as necessary [Fortier, 1987].

Simulations, beyond their structure (elements and interactions), require data input and data extraction to make them useful. The most usual simulations are either self-driven or trace-driven. In self-driven simulations the model itself (i.e., the program) has drivers embedded in it to provide the needed data to stimulate the simulation. This data is typically derived by various analytical distributions and linked with a random number generator. In the example of the bank teller system, we have been using a self-driven simulation. We may use a Poisson arrival distribution to describe the random nature of customers arriving to the system. Such a use is indicative of some artificially generated stream-to-model system inputs. In the other case, when we use trace-driven data, the simulation is being driven by outside stimuli. Typically this is extracted, reduced, and correlated data from an actual running system. For example, in our bank teller case we may wish to have a more realistic load base from which to compute the optimal number of tellers and their hours. In such a case we would measure over some period of time the dynamics of customers arriving to the bank for service. This collected information would then be used to build a stored input sequence that would drive the simulation based on this realistic data. This type of modeling is closer to the real world system but has the disadvantage of requiring the up-front data collection and analysis to make such data available for use.

TIME CONTROL

In continuous and discrete simulation, the major concern in performing the simulation is time management and its use in affecting the dependent variables. Timing in simulation programs is used to synchronize events, compute state changes, and control overall interactions. Timing can take on two modes: synchronous and asynchronous.

Synchronous timing refers to a timing scheme in which time advances in fixed, appropriately chosen units of time t. On each update of time the system state is updated to reflect this time change. That is, all events occurring during this time period are determined and their state adjusted accordingly. This process of advancing time (in steps) and updating the state of elements occurs until the simulation hits some boundary condition (time goes to a maximum, some event occurs, etc.). In our bank teller system timing is needed to determine arrivals and service. For the t-step organization on each stop we must check to see if an arrival should occur, if a service should be completed, or if a new one should be begun. An important concept or idea to keep in mind when using synchronous timing is that of step selection. If too great a step is chosen, events are seen to occur concurrently when in reality they may not. On the other hand, too fine a granularity of time step will cause many steps to go by when nothing occurs. The latter will cause excessive computer run time, but very fine differentiation between events. The former, on the other hand, will cause a fuzzing of everything and possibly a loss of usefulness. The important job of the modeler is to select the proper step time for the model to be useful, but not be excessive in computer time.

Asynchronous, or event timing, differs from synchronous timing in that time is advanced in variable rather than fixed amounts. The concept is to keep track of events versus time steps. The time is advanced based on the next chronological event that is to occur. These chronological events are typically kept in a dynamic list that is updated and adjusted on each event to reflect new events that are to occur in the future. In our bank teller example the event queue, or list, will comprise two events, the next arrival and the completion of the next service. Abstractly this method appears to be easier to visualize. The events must be ordered by occurrence and adjusted as new events arrive. The issue in this as well as in the former case is how to insert or "schedule" new events or new conditions in the simulations. The next section will investigate this and other aspects of how to use time in building simulations.

SYSTEMS AND MODELING

Up to this point, we have discussed generic attributes related to simulation modeling. We have not discussed the classes of modeling techniques available or the classification of simulation implementation techniques (i.e., simulation languages). Simulation techniques include discrete event, continuous change, queuing, combined, and hybrid techniques. Each provides a specific viewpoint to be applied to the simulation problem. They will also force the modeler to fit models to the idiosyncrasies of the techniques.

DISCRETE MODELS

In discrete simulation models, the real system's objects are referred to typically as entities. Entities carry with them attributes that describe them (i.e., their state description). Actions on these entities occur on boundary points or conditions. These conditions are referred to as events. Events such as arrivals, service stand points, stop points, other event signalling, wait times, etc., are typical.

The entities carry attributes that provide information on what to do with them based on other occurring events and conditions. Only on these event boundaries or condition occurrences can the state of entities change. For example, in our bank teller simulation, only on an arrival of a customer (arrival event) can a service event be scheduled or only on a service event can an end of service event be scheduled. This implies that without events the simulation does not do anything. This modeling technique only works on the notion of scheduling events and acting on them. Therefore, it is essential that the capability exists to place events into a schedule queue or list and to remove them based on some conditions of interest.

What this technique implies is that all actions within the simulation are driven by the event boundaries. That is, event beginnings and endings can be other events to be simulated (i.e., to be brought into action). All things in between these event boundaries, or data collection points, are now changing. A simulation model using this technique requires the modeler to define all possible events within the real system and how these events affect the state of all the other events in the system. This process includes defining the events and developing definitions of change to other states at all event boundaries, of all activities that the entities can perform, and of the interaction among all the entities within the simulated system. In this type of simulation modeling each event must trigger some other event within the system. If this condition does not hold, we cannot construct a realistic working simulation. This triggering provides the event's interaction and relationship with each other event. For example, for the model of a self-service automatic teller machine, we need to define at a minimum the following entities and events:

- Arrival events
- Service events
- Departure events
- Collection events
- Customer entities
- Server entities

The events guide how the process occurs and entities provide the media being acted on, all of which are overseen by the collection event that provides the "snapshot" view of the system. This provides a means to extract statistics from entities. In the above example, to build a simple model the following descriptions could be used:

1. Arrival event
   a. Schedule next arrival (present time + T)
   b. If all tellers busy, number waiting = number waiting + 1
   c. If any teller is free and no one before the waiting customer, schedule service event end
2. Service event
   a. Number of tellers busy = number tellers busy + 1
   b. Schedule service and event based on type of service
   c. Take begin service statistics

3. End service event
   a. Number tellers busy = number tellers busy - 1
   b. Schedule arrival of customer
   c. Take end of service statistics
4. Entities
   a. Tellers
      (1) Number of tellers
      (2) Service rates and types
      (3) Service types
   b. Customers
      (1) Arrival rate
      (2) Dynamics (service type required)

A discrete event simulation (with an appropriate language) could be built using these events and entities as their basis. A model built this way uses these conditions to schedule some number of arrivals and some end conditions. The relationships that exist between the entities will keep the model executing with statistics taken until an end condition is next. This example is extremely simplistic and by no means complete, but it does provide a description of some of the basic concepts associated with discrete event simulations. Details can be found in [Pritsker, 1984] and its references.

CONTINUOUS MODELING

Continuous simulations deal with the modeling of physical events (processes, behaviors, conditions) that can be described by some set of continuously changing dependent variables. These in turn are incorporated into differential, or difference, equations that describe the physical process. For example, we may wish to determine the rate of change of speed of a falling object shot from a catapult (see Figure 5-1) and its distance R. The equations for this are as follows. The velocity v at any time is found as

$$\frac{d_v}{d_t} = v_x^2 = v_y^2$$

and

$$v_x = v_o \cos 0 \quad v_y = v_o \sin 0_o - g_t$$

and the distance is

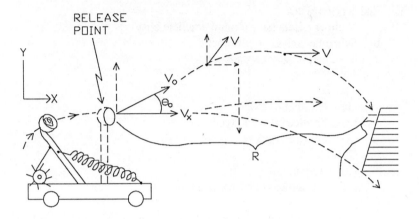

Figure 5-1 Projectile motion.

$$R = \frac{\tan \theta_0}{2CV_o \cos \theta_o}$$

and the distance at any time

$$\frac{d_y}{d_t} = (Vp_{v_o} \, P_{v_{\cos\theta}})^2$$

These quantities can be formulated into equations that can be modeled in a continuous language to determine their state at any period of time t.

Using these state equations, we can build state-based changed simulations that provide us with the means to trigger on certain occurrences. For example, in the above equations we may wish to trigger an event (shoot back when v_y is 0). That is, when the projectile is not climbing any more and it has reached its maximum height, fire back. In this event the code may look like this:

when $v_y = 0$; begin execution of shootback

Another example of this type of triggering is shown in Figure 5-2. In this example, two continuous formulas are being computed over time; when their results are equivalent (crossover event), schedule some other event to occur. This type of operation allows us to trigger new computations or adjust values of present ones based on the relationship of continuous equations with each other.

Using combinations of self-triggers and comparative triggers (less than, greater than, equal to, etc.) we can construct ever more involved simulations of complex systems. The main job of a simulator in this type of simulation model is to develop a set of equations that define the dynamics of the system under study and determine how they are to interact.

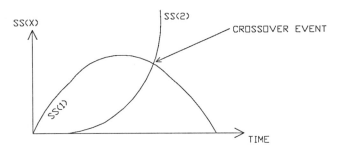

Figure 5-2 Continuous variable plot.

QUEUING MODELING

Another class of generic model is the queuing model. Queuing-based simulation languages exist (Q-gert, Slam) and have been used to solve a variety of problems. As was indicated earlier, many problems to be modeled can easily be described as an interconnection of queues, with various queuing disciplines and service rates. As such, a simulation language that supports queuing and analysis of them would greatly simplify the modeling problem. In such languages there are facilities to support the definition of queues in terms of size of queue, number of servers, type of queue, queue discipline, server type, server discipline, creation of customers, monitoring of operations, departure collection point, statistics collection and correlation, and presentation of operations. In addition to basic services there may be others for slowing up customers or routing them to various places in the queuing network. Details of such a modeling tool will be highlighted in the later sections of this chapter.

COMBINED MODELING

Each of the above techniques provides the modeler with a particular view upon which to fit the system's model. The discrete event-driven models provide us with a view in which systems are composed of entities and events that occur to change the state of these entities. Continuous models provide a means to perform simulations based on differential equations or difference formulas that describe time-varying dynamics of a system's operation. Queuing modeling provides the modeler with a view of systems being comprised of queues and services. The structure comes from how they are interconnected and how these interconnections are driven by the outputs of the queue servers.

The problem with all three techniques is that in order to use them, a modeler must formulate the problem in terms of the available structure of the technique. It cannot be formulated in a natural way and then translated easily. The burden of fitting it into a framework lies on the modeler and the simulation language. The solution is to provide a combined language that has the features of all three techniques. In such a language the modeler can build simulations in a top-down fashion, leaving details to latter levels. For instance, in our early bank teller system, we could initially model it as a single queue with n servers (tellers). The queuing discipline is first come, first served and the service discipline can be any simple distribution such as exponential. This simple model will provide us with a sanity check of the correctness of our model and with bounds to quickly determine the system's limits. We could next decide to model the teller's service in greater detail by dropping this component's level down to the event modeling level.

At this point we could model the teller's activity as a collection of events that need to be sequenced through in order for service to be completed. If possible, we could then incorporate continuous model aspects to get further refinement of some other feature. The main aspect to gather from this form of modeling is that it provides the modeler with the ability to easily model the level of detail necessary to simulate the system under study.

HYBRID MODELING

Hybrid modeling refers to simulation modeling in which we incorporate features of the previous techniques with conventional programming languages. This form of modeling could be as simple as doing the whole thing in a regular language and allowing lower levels of modeling by providing a conventional language interface. Most simulation languages provide means to insert regular programming language code into them. Therefore, they all could be considered a variant of this technique.

SIMULATION LANGUAGES

As the use of simulation has increased, so has the development of new simulation languages. Simulation languages have been developed because of the unique needs of the modeling community to have system routines to keep track of time, maintain the state of the simulation, collect statistics, provide stimulus, and control interaction. All of these previously had to be done by each individual programmer.

Early languages provided basic services by adding a callable routine from programming languages. These early languages provided for time and event management but little else. This chapter will look at five languages and discuss the aspects that they possess that aid in the simulation process. It will not, however, cover languages that are built on top of basic simulation languages such as Network II.5 and others.

GASP IV

GASP IV was developed in the early 1970s as a general-purpose simulation language. It is a Fortran-based simulation language that provides routines and structure to support the writing of discrete event, continuous and combined discrete event, and continuous simulation models. Discrete event models in GASP IV are written as a collection of system and user Fortran subroutines. GASP IV provides the user with the following routines: time management, file management (event files, storage and retrieval, copying of events, and finding of events), and data collection and analysis (both observation based and time based statistics). The user must develop a set of event routines that define and describe the mathematical-logical relationships for modeling the changes in state corresponding to each event and their interactions.

As an example of GASP IV's use and structure, our bank teller problem will be examined once again. In order to model this problem in GASP we must determine the events of interest, their structure, and the boundaries upon which they are triggered. To simplify the example, it is assumed that there is no time delay between the ending of service for one customer and the beginning of another (if there is one waiting). The important measures or states will be the number of customers in the system and the teller's status. From these two system events arise a customer's arrival and a teller's end of service. These are also chosen as the points at which significant changes to a system's status occur. The activity that occurs is the beginning of service; this can be assumed to occur either when a customer arrives at an empty line or when the teller ends service to a customer.

Entities in GASP are represented by arrays of attributes where the attributes represent the descriptive information about the entity that the modeler wishes to keep. Entities are the elements that are acted on during the simulation. Their attributes are adjusted based on occurrences of interest. A variable "busy" is used to indicate the status of the teller, and attribute (1) of customer is used to mark the customer's arrival time to the teller line. To make the simulation operate, the system state must be initialized to some known values; in this case the teller is initialized not busy and the first arriving customer must be scheduled to arrive. Additionally, to keep the model running, the arriving customer must schedule another customer's arrival in the future based on a selected random time distribution. Statistics will be taken when service completes on the length of wait time and the number of customers waiting, in service, and in total. When we look at the GASP code we need to examine the structure of a typical GASP program (see Figure 5-3). As indicated by this figure, GASP IV exists as a single program in Fortran. Therefore, making it function requires a main program and a call to the GASP program that will begin the simulation. Another function of the main program is to set up limits on the system such as number of files, input, output, limits on events, etc. Figure 5-4 depicts the main program for our example.

Once GASP has been called, the program runs under control of the GASP subprogram calling sequences. The GASP system's executive takes over and initializes the simulation using the user-supplied routine Intlc. Once initialized, it begins simulation by examining the event list; if any exists, it pulls them out, executes them, and takes statistics. The loop continues until the end conditions are met or an error occurs. To control events and make sense of them in Fortran requires the user to supply an event-sequencing routine called Event. This event control

96 Modeling and Analysis of Local Area Networks

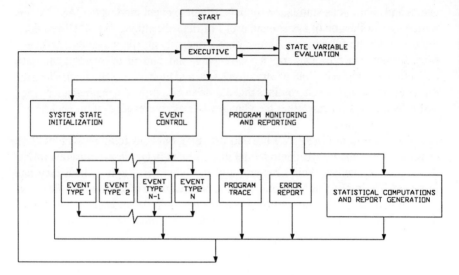

Figure 5-3 Basic model of GASP IV control.

routine is called with the attribute number of the intended event. It will use this to call the actual event. For our bank teller example this routine is illustrated in Figure 5-5 with its two events shown. When this routine is called with an appropriate number, the intended event is called, executed, and control is returned to the event routine, which in turn returns control to the executive routine.

These events are called based on what filem (1) has stored in it. Filem (1) is operated on in a first come, first serve basis, removing items one at a time. The events are stored in filem (1) as attributes as follows: attribute (1) is the time of the

```
Dimension nset (1000)
common/gcom1/atrib (100, DP(100),DDL (100, DTNOW,
II,MFA,MSTOP,NCLNR,NCRDR,NPRINT,NNRUN,
NNSET,NTAPE,SS(100),SSL(100),TNEXT,TNOW,XX(100)
COMMON Q Set
Equivalence(NSet(1),QSet(1))
NNSet=1000
NCRDR=5
NPRINT=6
NTAPE=7
CALL GASP
STOP
END
```

Figure 5-4 GASP main Fortran program.

```
Subroutine Event (I)
Go to (1,2),I
1 Call arrival
  return
2 Call end SRU
  return
  end
```

Figure 5-5 Subroutine event for bank teller.

event, attribute (2) is the event type, and all other attributes are added user attributes for the entity. Figure 5-6 gives an example of how the file is initialized. This figure illustrates the initialization routine for the bank teller simulation. Filem (1), the event file, is loaded with the first arrival event (a customer) and the teller is set to not busy.

Once filem (1) has an event stored, the simulation can begin. The first event is a customer arrival indicated by the contents of attribute (2) of filem (1), which is the only event at this time. The arrival event (see Figure 5-7) performs the scheduling of the next arrival.

After the next arrival is scheduled, the preset arrival is placed in the queue [filem (2)]. Then a test is made to see if the teller is busy. If so, we return to the main program or else we schedule an end of service event for the preset time plus a number chosen uniformly between 10 and 25.

The second event, the end of service, is shown in Figure 5-8. This code determines statistics of time in system and busy statistics. The code also checks to see if there is any user in the queue, removes one if there is, and schedules another end of service for this user.

This simple example shows how GASP could be used to model a wide array of event-based models. Details of this language can be found in [Pritsher, 1974].

```
Subroutine INTLc
common/gcoml/attrib(100),DD(100),DDL(100),DTNOW,D
Mfa/mstop,nclnr,ncrdr,nprant,nnrun,nnset,ntape,
ss(100),ssl(100),tnext,tnow,xx(100
Equivalence (xx(1),Busy)
Busy=0.
Atrib(2)=1
call filem
 return
 end
```

Figure 5-6 Subroutine Intlc for bank teller.

```
Subroutine Arrival
   common/gcoml/atrib(100,DD(100),DDL(100),DTNOW,II,MFA
   SSL(100),TNEXT,TNOW,XX(100)
   Equivalence(XX(1),BUSY)TIMST(BUSY,TNOW,ISTAT)\
Atrib(1)=Tnow+expon(20.,1)
   Atrib(2) = 1
   Call filem(1)
   Call filem (2,attrib(1))
   if (busy,e??q.0) go to
   return
10 busy =1.
   attrib(1)=tnow+unfrm (10., 25, 1)
   attrib:(2) = 1
   attrib(3) = tnow
   call filem (1)
   return
   end
```

Figure 5-7 Arrival event code.

GPSS

General-Purpose Simulation System (GPSS) is a process-oriented simulation language for modeling discrete systems. It uses a block-structuring notation to build models. These provide a set of standard blocks (see Figure 5-9) that provide the control and operations for transactions (entities). A model is translated into a GPSS

```
Subroutine end SRU
common/Gcoml/atrib(100),DD(100),DD(100),DTN00,
II,MFA,MSTOP,NCLNR,NCRDR,NPRNT,NNRUN,NNSET,NTAPE
SS(100),SSL(100),TNEXT,TNOW,XX(100)
Timst (subusyTNOW,T
ISTAT)
   TSYS=TNOW_Atrib(3)
   Call col Ct(Tsys,1)
   if(nnQ(2),6T,0)go to 10
   busys=0
   return
   10 call remove (1,1,atrib(3))
   call schd((2,unfrm(10.,25.,1)atrib)
   return
   end
```

Figure 5-8 For server routine.

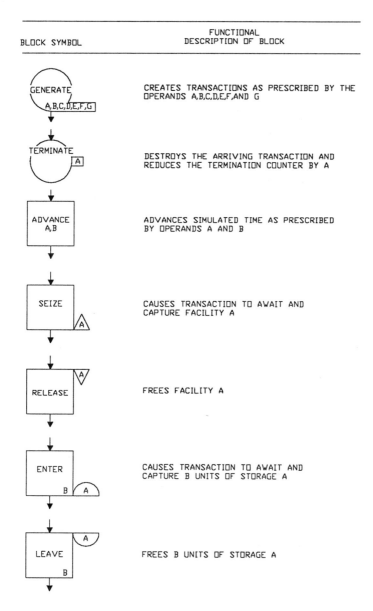

Figure 5-9(a)   Basic GPSS blocks.

program by the selection of blocks to represent the model's components and the linkage of them into a block diagram defining the logical structures of the system. GPSS interprets and executes the block diagram defined by the user, thereby providing the simulation. This interpretation is slow and, therefore, the language cannot be used to solve large problems.

BLOCK SYMBOL	FUNCTIONAL DESCRIPTION OF BLOCK
QUEUE A, B	INCREMENTS THE NUMBER IN QUEUE A BY B UNITS
DEPART A, B	DECREMENTS THE NUMBER IN QUEUE A BY B UNITS
A,B,C ASSIGN	ASSIGNS THE VALUE SPECIFIED AS B WITH MODIFIER C TO PARAMETER NUMBER A OF THE TRANSACTION
MARK A	ASSIGNS THE CURRENT CLOCK TIME TO PARAMETER NUMBER A OF THE TRANSACTION
SAVEVALUE A,B,C	ASSIGNS THE VALUE SPECIFIED AS B TO SAVEVALUE LOCATION A
TRANSFER A (C) (B)	CAUSES A TRANSFER TO LOCATION C WITH PROBABILITY A, AND LOCATION B WITH PROBABILITY 1-A
TEST X A B (C)	CAUSES A TRANSFER TO LOCATION C IF A IS NOT RELATED TO B ACCORDING TO OPERATOR X
TABULATE A	RECORDS AN OBSERVATION FOR THE VARIABLE PRESCRIBED IN TABLE A

Figure 5-9(b)   Basic GPSS blocks.

To illustrate GPSS we will again examine our bank teller system. It is viewed as a single-server queuing system with our teller and n customer arrivals (see Figure 5-10). Customers arrive with a mean interarrival time of 10 minutes, exponentially distributed. The teller provides service to the customers in a uniform time between 5 and 15 minutes. The simulation will take statistics on queue length, utilization of

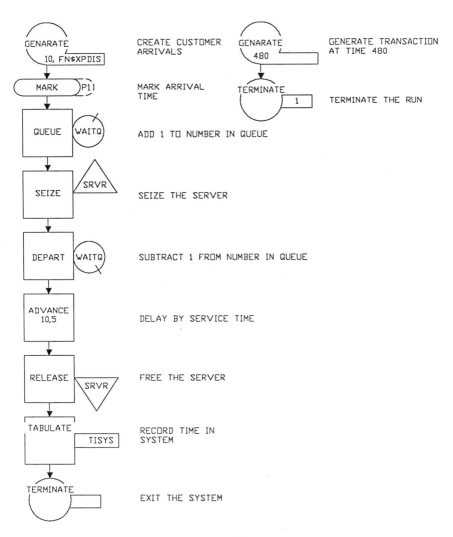

Figure 5-10 GPSS model for bank teller problem.

teller, and time in system. The simulator builds the diagram shown in Figure 5-10 from the model of a queuing system and based on the statistics to be taken. This structure is then translated to the code seen in Figure 5-11. The code is broken down into four sections. The top section is used to define the data needed to approximate an exponential distribution and set up markers for the time-dependent statistics.

The second segment is the main simulation code, and it performs the tasks of generating customers (14), taking statistics on arrival time (15), queuing up arriving customers (16), scheduling service (17; when free, take control), departing the

```
 1  *
 2  SIMULATE
 3  *
 4  XPDIS FUNCTION RN1,C24
 5  0.0,0.0/0.1,0.104/0.2,0.222/0.3,0355/0.4,0.509/0.5,0.69
 6  0.6,,0.915/0.7,1.2/0.75,1.38/0.8,1.6/0.84,1.83/0.88,2.12
 7  0.9,2.3/0.92,2.52/0.94,2.81/0.95,2.99/0.96,3.2/0.97,3.5
 8  0.98,4.0/0.99,4.6/0.995,5.3/0.998,6.2/0.999,7/0.9997,8
 9  *
10  TISYS TABLE MP1,0,5,20
11  *
12  * MODEL SEGMENT
13  *
14  GENERATE 10,FN$XPDIS
15  MARK P1,
16  QUEUE WAITQ
17  SEIZE SRVR
18  DEPART WAITQ
19  ADVANCE 10,5
20  RELEASE SRVR
21  TABULATE TISYS
22  TERMINATE
23  *
24  * TIMING SEGMENT
25  *
26  * GENERATE 480
27  TERMINATE 1
28  *
29  * CONTROL CARDS
30  * 31 START 1 32 END
```

Figure 5-11 GPSS coding for bank teller problem.

waiting line (18), delaying the exit by the appropriate service time of the teller (19), releasing the teller for the next customer (20), taking statistics on the customer's time in the system (21), and exiting the system (27).

The third segment is a timing segment and is used to schedule the end of service routine. The model will schedule a dummy transaction at time 480, which will cause the terminate instruction to execute (counter set to 0). The fourth section, the control segment, begins the simulation by setting the termination counter and giving control over to the model segment.

This example shows some features of GPSS; refer to [Schriber, 1974] for details of this language. GPSS is a simple modeling method that became widely used. However, this language was doomed by its interpretive operation, which made it extremely slow.

SIMSCRIPT

Simscript was developed in the late 1960s as a general-purpose simulation language [Rivait, 1969]. It provides a discrete simulation modeling framework with English-like free-form syntax making for very readable and self-documenting models. Simscript supports two types of entities, permanent and temporary. For example, in the bank teller problem, the teller is permanent and the clients are temporary. Permanent entities exist for the entire duration of the simulation, whereas the temporary entities come and go during it. Attributes of the entities are named, increasing their readability and meanings.

A Simscript simulation is built of three pieces: a preamble, a main program, and event subprograms. The preamble defines the components of the model (entities, variables, arrays, etc.). The main program initializes all elements to begin the simulation. The events define the user events used to model a system. To define these components we will again use the bank teller problem defined earlier. We will assume arrivals are 10 minutes apart on average and exponentially distributed, and the teller service time is uniformly distributed between 5 and 15 minutes. Figure 5-12 depicts code for this problem. It indicates many of Simscript's features; for instance:

- Line 2 describes the wait time as being a system entity that has statistics associated with it, and it is a permanent entity since it is not indicated as being temporary. Therefore, we can keep statistics on it over the life of the model.
- Line 4 defines a temporary entity customer and indicates that it belongs to the wait time.
- Lines 5 and 6 define the event names and their attributes.
- Lines 12, 13, 14, and 15 define statistics to be taken on this entity.

The main program or section is shown in section B. This portion sets up the initial conditions (i.e., setting the status of the teller to idle, scheduling the first arrival, and scheduling a stop in the simulation). The next three sections define the arrival, departure, and stop events. The arrival event schedules the next arrival to keep the event flow going, creates a customer, gives it time information, places it in the wait line, and schedules a teller service if the line is empty. The departure event computes a customer's time in the bank, removes the customer, and schedules the next customer. The stop event outputs the collected statistics. Details of this language can be found in the previously provided reference.

SLAM II

Slam II, simulation language for alternative modeling, was developed by Pritsker and Associates, West Lafayette, Indiana, in the late 1970s. It is a combined modeling language providing for queuing network analysis, discrete event, and continuous modeling in integrated form. Slam II provides features to easily integrate the three forms. At the highest end the modeler can use a network structure consisting of nodes and branches representing queues, servers, and decision points to construct a model of a system to be simulated. This, in turn, can be easily translated into Slam II code. Additionally, Slam provides the ability to mix events

```
A:
1  PREAMBLE
2  THE SYSTEM OWNS A WAIT.LINE AND HAS A STATUS
   TEMPORARY ENTITIES
4  EVERY CUSTOMER HAS A ENTER.TIME AND MAY BELONG TO THE
WAIT.LINE
5  EVENT NOTICES INCLUDE ARRIVAL AND STOP.SIMULATION
6  EVERY DEPARTURE HAS A TELLER
7  DEFINE BUSY TO MEAN 1
8  DEFINE IDLE TO MEAN 0
9  DEFINE TIME.IN.BANK AS A REAL VARIABLE
10 TALLY NO.CUSTOMERS AS THE NUMBER,AV.TIME AND THE MEAN,
11 AND VAR.TIME AS THE VARIANCE OF TIME.IN.BANK
12 ACCUMULATE AVG.UTIL AS THE MEAN,AND VAR.UTIL AS THE
13 VARIANCE OF STATUS
14 ACCUMULATE AVE.WAITLINE.LENGTH AS THE MEAN, AND
15 VAR.WAITLINE.LENGTH AS THE VARIANCE OF N.WAIT.LINE
16 END

B:
1  MAIN
2  LET STATUS=IDLE
3  SCHEDULE AN ARRIVAL NOW
4  SCHEDULE A STOP.SIMULATION IN 8 HOURS
5  START SIMULATION
6  END

C:
1  EVENT ARRIVAL
2  SCHEDULE AN ARRIVAL IN EXPONENTIAL.F(10.,1) MINUTES
3  CREATE A CUSTOMER
4  LET ENTER.TIME(CUSTOMER)=TIME.V
5  IF STATUS=BUSY
6  FILE THE CUSTOMER IN THE WAIT.LINE
7  RETURN
8  ELSE
9  LET STATUS=BUSY
10 SCHEDULE A DEPARTURE GIVEN CUSTOMER IN UNIFORM.F(5.,15.,
11 1) MINUTES
12 RETURN
13 END
```

Figure 5-12 Simscript bankteller code.

```
D:

 1 EVENT DEPARTURE GIVEN CUSTOMER
 2 DEFINE CUSTOMER AS AN INTEGER VARIABLE
 3 LET TIME.IN.BACK=1440.*(TIME.V-ENTER.TIME(CUSTOMER))
 4 DESTROY THE CUSTOMER
 5 IF THE WAIT.LINE IS EMPTY
 6 LET STATUS=IDLE
 7 RETURN
 8 ELSE
 9 REMOVE THE FIRST CUSTOMER FROM THE WAIT.LINE
10 SCHEDULE A DEPARTURE GIVEN CUSTOMER IN UNIFORM.F(5.,
11 15.,1) MINUTES
12 RETURN
13 END

E:
 1 EVENT STOP.SIMULATION
 2 START NEW PAGE
 3 SKIP 5 LINES
 4 PRINT 1 LINE THUS
   SINGLE TELLE RWAIT.LINE EXAMPLE
 6 SKIP 4 LINES
 7 PRINT 3 LINES WITH NO.CUSTOMERS, AV.TIME, AND VAR.TIME
   THUS
NUMBER OF CUSTOMERS = *********
AVERAGE TIME IN BANK = ****.****
VARIANCE OF TIME IN BANK = ****.****
11 SKIP 4 LINES
12 PRINT 2 LINES WITH AVG.UTIL AND VAR.UTIL THUS
AVERAGE TELLER UTILIZATION = ****.****
VARIANCE OF UTILIZATION = ****.****
15 SKIP 4 LINES
16 PRINT 2 LINES WITH AVE.QUEUE.LENGTH AND VAR.QUEUE.LENGTH
   THUS
AVERAGE WAIT.LINE LENGTH = ****.****
VARIANCE OF WAIT.LINE = ****.****
19 STOP
20 END
```

Figure 5-12 (cont.)   Simscript bankteller code.

and continuous models with network models by use of event nodes that call event code for discrete and/or continuous models. As in the previous languages, the event-oriented Slam models are constructed of a set of events and the potential changes that can occur with each of them. These events define how the model interprets the event and state changes. Slam provides a set of standard support subprograms to aid the event-oriented modeler. As was the case in GASP, the Slam continuous models are built by specifying a set of continuous differential, or difference, equations that describe the dynamic behavior of the state variables. These equations are coded in Fortran (Slam's base language) using Slam's state variables.

Slam II uses a set of basic symbols to describe the system being modeled, as does GPSS. Figure 5-13 depicts the basic Slam II symbols and their associated code

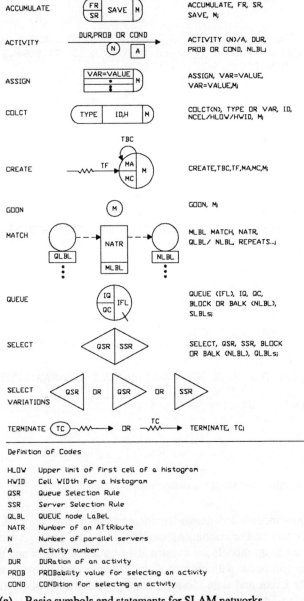

Figure 5-13(a) Basic symbols and statements for SLAM networks.

Simulation Analysis   107

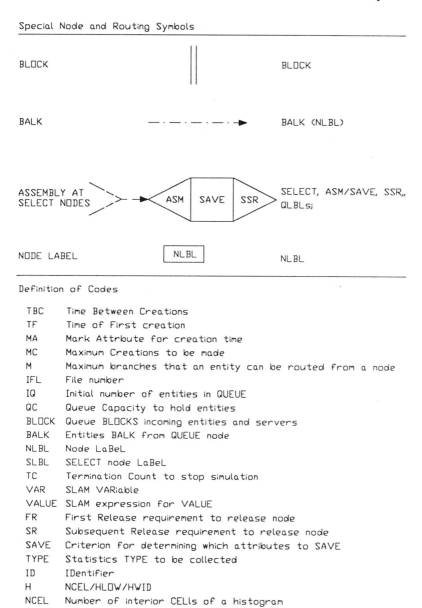

Figure 5-13(b)   Basic symbols and statements for SLAM networks.

statements. Only the first three characters of the statement names and the first four characters of node labels are significant. They will be used in the example of the bank teller. As before, we wish to have customers arriving on an average of every 10 minutes with an exponential distribution and the first one to start at time 0. Additionally, the teller services the customers with a uniform distribution from 5 to 15 minutes. The resulting Slam network model is shown in Figure 5-14.

108 Modeling and Analysis of Local Area Networks

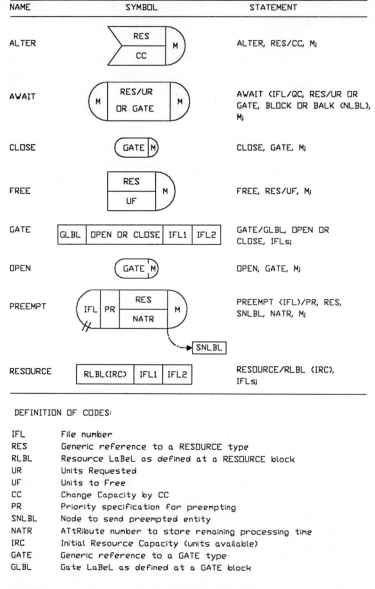

Figure 5-13(c) Basic symbols and statements for SLAM networks.

References to nodes are made through node labels (NLBLs). When a node label is required, it is placed in a rectangle and appended to the base of the symbol.

The code for this network is shown in Figure 5-15. The first line of the code defines the modeler, the name of the model, and its date and version. The second line defines the limits of the model and files one USR attribute and up to 100

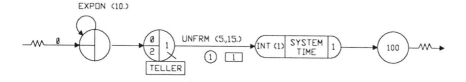

Figure 5-14 Bank teller network model.

concurrent entities in the system at a time. Line 3 identifies this code as network code, and line 4 creates customers with a mean of 10 minutes exponentially distributed. Line 5 defines queue 1 as a teller with no initial customers in its queue, an infinite queue with service uniformly distributed from 5 to 15 minutes. Line 7 takes statistics or time in system from entities as they leave the server. Line 8 indicates that the simulation will run for 100 entities and then end.

This code is extremely simple and provides much flexibility as to how to expand the system. To look at the tellers' operations in more detail, the queue could be replaced by an event node and the code for the teller event supplied to model (very similar to the code seen in earlier figures). Details of Slam II and its uses can be found in [Pritsker, 1984].

APPLICATIONS OF SIMULATION

To illustrate the use of simulation a few example problems are given and models developed in the Slam II simulation language. The first example is an industrial plant with five stations building a production in assembly line fashion. Pictorially the problem can be viewed as in Figure 5-16.

The plant takes in subassemblies and finishes them off in five steps. There is storage room at the beginning of the line, but once in the line a maximum of one unit per station is possible. The statistics we wish to determine are workstation utilization, time to process through stations, number of units waiting, and total produced. The resulting Slam network is shown in Figure 5-17. The resultant code

```
1       GEN, FORTIER,BANKTELLER,8/12/88,1;
2       LIMITS,2,1,100
3       NETWORK;
4            CREATE,EXPON(10.),0;1;
5  TELLER QUEUE(1),0,~;
6            ACTIVITY (1)/1,UNIFRM(5.,15.);
7            TERM 100; COLCT,INI(1),SYSTEM TIME,,1;
8            END NETWORKS
```

Figure 5-15 SLAM bank teller code.

110 Modeling and Analysis of Local Area Networks

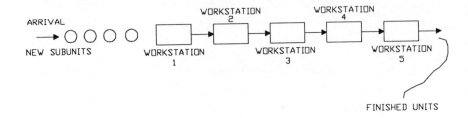

Figure 5-16 Assembly line.

would allow us to examine the items of interest without causing any loss of detail from the intended model.

A second more detailed example shows how simulation can be used to model a distributed database management system. The model is shown in Figure 5-18. Depicted is the process or servers in a node that services user database transactions. Users provide requests, the operating system services them by pipelining the database ones to the transaction manager, which in turn provides reduced requests to the network database server, who determines where the actual access is to be performed. The local site chosen then accesses the information from the appropriate device. The details at each level were commensurate with the intended model. The queues were all modeled as events and then the code necessary to simulate them was developed. This simulation is being used to analyze optimization algorithms for distributed database systems. The Slam network and its code are shown in Figures 5-19 and 5-20. Details of this model can be found in [Fortier, 1986].

THE SIMULATION PROGRAM

The simulation program constitutes the realization of the simulation model. It is constructed as a modular software package allowing for the interchanging of simulated database management components without causing undue stress to the other components of the model. The simulation program is composed of a set of Slam II network statements and detailed discrete event code (similar to GASP IV),

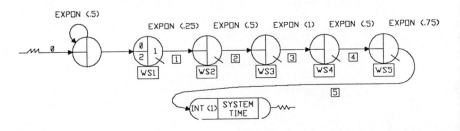

Figure 5-17 SLAM II network.

Simulation Analysis   111

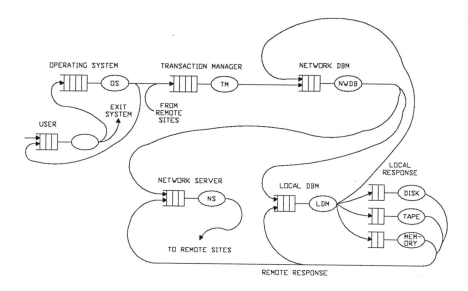

Figure 5-18 Queueing model of distributed database system.

which model the major computational aspects of a distributed database management system as previously defined. To provide the capability to model a wide range of topologies and database management architectures, the model is driven by a set of information tables that hold characteristics of the network topology and com-

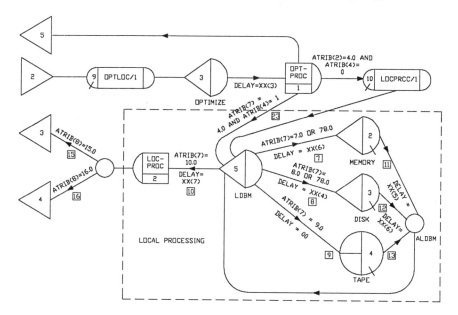

Figure 5-19(a) SLAM simulation model.

112 Modeling and Analysis of Local Area Networks

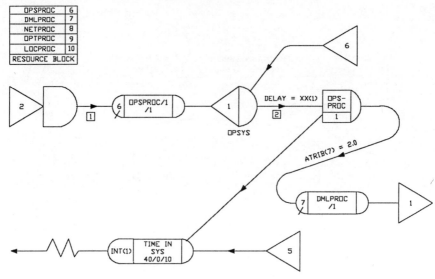

Figure 5-19(b) SLAM simulation model.

munications, location of the data items, the contents of the data items, and statistics of usage. The Slam II network code to realize this model is shown in Figure 5-20. This code clearly depicts the major components of the simulation program. Addi-

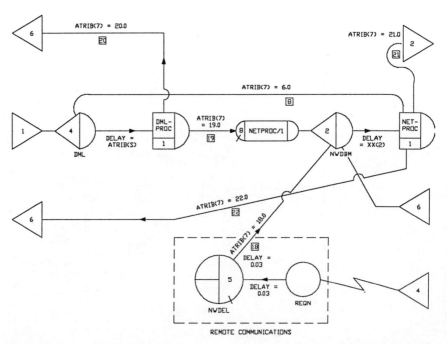

Figure 5-19(c) SLAM simulation model.

This is a list of the activities in the network:

```
1  ENTER to OPSYS    2  OPSYS to DML
3  OPSYS to USER    4  OPTIMIZE to LDBM
5  OPTIMIZE to USER  6  NWDBM to DML
7  LDBM to MEMORY   8  LDBM to DISK
9  LDBM to TAPE    10  LDBM to DLOC
11 MEMORY to RLDBM 12  DISK to RLDBM
13 TAPE to RLDBM   14  RLDBM to LDBM
15 DLOC to NWDBM   16  DLOC to REQN
17 REQN to NWDEL   18  NWDEL to NWDBM
19 DML to NWDBM    20  DML to OPSYS
21 NWDBM to OPTIMIZE 22 NWDBM to USER
23 LDBM to REQN    24  NWDBM to USER
```

The following statements are network input statements:

```
GEN,P. FORTIER, DBMS QUEUE SIMPROG, 7/12/84,1;
LIMITS,10,20,500;
STAT,1,HITS ON DIRECTORY,10,1.,1.;
STAT,2,HITS ON DICTNARY,10,1.,1.;
STAT,3,PROCESSING TIME,20,0.,10.;
STAT,4,REMOTE TIME,10,0.,.05;
STAT,5,FAILURE RATE,10,1.,1.;
STAT,6,OPTIMIZER TIME,10,0.,10.;
STAT,7,OPT. ALGO. DELAY, 10,0.,10.;
STAT,8,PARSING DELAY,10,0.,.0015;
STAT,9,ILLEGAL OPER.,10,1.,1.;
STAT,11,TRANSLATE DELAY,10,0.,.01;
STAT,12,DICTNARY SEARCH,10,0.,.00002;
NETWORK;
RESOURCE/OPSPROC(1),6;
RESOURCE/DMLPROC(1),7;
RESOURCE/NETPROC(1),8;
RESOURCE/OPTPROC(1),9;
RESOURCE/LOCPROC(1),10;
ENTER,1;
ACT/1;
OPSYS AWAIT(6),OPSPROC/1;
EVENT,1; OPERATING SYSTEM
ACT,XX(1);
FREE,OPSPROC/1;
ACT/3,,ATRIB(7).EQ.3,USER; SERVICE COMPLETED
```

Figure 5-20 SLAM II network code for database model.

```
ACT/2,,ATRIB(7),EQ.2;
DML AWAIT(7),DMLPROC/1;
EVENT,4;
ACT,ATRIB(5);
FREE,DMLPROC/1;
ACT/20,,ATRIB(7).EQ.20,OPSYS SERVICE COMPLETED
ACT/19,,ATRIB(7).EQ.19,NWDBM;
NWDBM AWAIT(8),NETPROC/1;
EVENT,2; NETWORK DATABASE MGR
ACT,XX(2);
FREE,NETPROC/1;
ACT/6,,ATRIB(7).EQ.6,DML PROCESSING COMPLETED
ACT/22,,ATRIB(7).EQ.22,USER; DATA DOESN'T EXIST
ACT/21,,ATRIB(7).EQ.21;
OPTIM AWAIT(9),OPTPROC/1;
EVENT,3; QUERY OPTIMIZATION
ACT,XX(3);
FREE,OPTPROC/1;
ACT/4,,ATRIB(7),EQ.4.0.AND.ATRIB(4).EQ.0,WLDBM;
ACT/26,ATRIB(7),EQ.4.0AND.ATRIB(4).EQ.1,LDBM;
ACT/5,,ATRIB(7).EQ.5,USER; ILLEGAL QUERY
WLDBM AWAIT(10),LOCPROC/1;
LDBM EVENT,5; LOCAL DATABASE MGR
ACT/7,XX(4),ATRIB(7),EQ.7.0.0OR.ATRIB(7),EQ.78.0,MEM;
ACT/8,XX(4),ATRIB(7),EQ.8.0.0OR.ATRIB(7),EQ.78.0.DISK;
ACT/9,999999,ATRIB(7),EQ.10.0,DLOC;
ACT/23,,ATRIB(7),EQ.23.0,REQN;
TAPE QUEUE(4);
ACT/13,,,RLDBM;
MEM QUEUE(2);
ACT/11,XX(5),,RLDBM;
DISK QUEUE(3);
ACT/12,XX(6)..RLDBM;
RLDBM GOON; REQUEST LDBM
ACT/14,,,LDBM;
DLOC FREE,LOCPROC/1;
ACT;
GOON;
ACT/15,,ATRIB(8).EQ.15,NWDBM RET.ROUTE,LOCAL SOURCE
ACT/16,0.02,ATRIB(8).EQ.16; RET. ROUTE, REMOTE SOURCE
REQN GOON;
ACT/17,0.02;
NWDEL QUEUE(5); NETWORK DELAY
ACT/18,0.03..NWDBM
USER COLCT,INT(1),TIME IN SYS,40,0.,10.;
```

Figure 5-20 (cont.)

```
ACT/20;
TERMINATE;
ENDNETWORK;
INIT,0;
FIN;
```

Figure 5-20 (cont.) SLAM II network code for database model.

tionally, note that the EVENTS,X shown indicates that the particular node is not a simple queue representation and also indicates a drop in detail into discrete event simulation code. Such events allow for greatly expanding the details of the aspect of the model.

SUMMARY

This chapter introduced the use of simulation in building and analyzing a wide range of systems. Simulations were shown to be extremely versatile in their ability to model systems at varying levels of detail. They provide quick and precise models of systems to allow any studies to be performed at will. The main simulation techniques of discrete event, continuous, queuing, combined, and hybrid methods were described, as were four well-used languages: GASP, GPSS, Simscript, and Slam. This was followed by two simple examples to show how simulation can be used to study a real world system.

REFERENCES

Fortier, Paul J. *Design of Distributed Operating Systems*, McGraw-Hill, 1986.

Fortier, Paul J. Handbook of LAN Technology, McGraw-Hill, 1989.

Pritsker, A. A. B. *Introduction to Simulation and SLAM II*, System Pub. Company, 1984.

Pritsker, A. A. B. *The GASP IV Simulation Language*, Wiley and Sons, Inc., 1974.

Krivait, J. *Simscript II Programming Language*, Prentice-Hall, 1969.

Schriber, A. *Simulation Using GPSS*, John Wiley, 1974.

6

Queuing Theory

In this chapter, we will build upon the basic probability theory covered in Chapter 4. The discussions will lead to the definition and analysis of several useful queuing models for the behavior of many types of service systems. The methods discussed herein complement those provided by simulation analysis. Frequently, the development of a general queuing model for a particular system will aid in the development of a detailed simulation of specific parts of the system. Also, the results and behavior observed from simulation help to tune the analytical models.

This chapter is organized into three general topics, stochastic process, queuing models, and estimation. Stochastic processes form the basis for many of the analytic techniques that apply to the queuing systems that we will discuss. The section on estimation provides some methods for defining the values that parameterize the queuing models with real world data.

STOCHASTIC PROCESSES

Although the probability distribution functions presented in the beginning of Chapter 4 are representative of certain real world phenomena, they alone cannot describe processes whose characteristics change over time. Thus, we may in fact end up with a family of random variables, each of which represents the service time characteristics for a given period. This family of random variables that collectively describe a time-varying process is known as a "stochastic process." A stochastic process is defined as

$$\{X(t), t \, \varepsilon \, T\} \tag{6-1}$$

where $X(t)$ is a random variable defined by the parameter t over the set T. The values that $X(t)$ may assume form the state space of the process and each value in the state space is known as a state of the random variable. As with random variables, stochastic processes are classified as continuous or discrete. If the number of states in the state space is finite, the process is discrete space. An infinite number of states denotes a continuous process. Discrete stochastic processes whose state space is the set {0, 1, 2,...} are also referred to as "chains." Continuous state spaces give rise to continuous-state processes. The type of parameter, t, that the stochastic process is defined over also defines a characteristic of the process. A discrete parameter index yields a discrete-parameter process while a continuous index yields a continuous-parameter process. In many cases, the index, also referred to here as the parameter, is time. Table 6-1 summarizes the state and parameter relationships.

A stochastic process is said to be stationary in the strict sense if all of the process statistics (e.g., mean, standard deviation, etc.) are independent of a shift in the origin of the process parameter. That is, if we look at a process at time t and note its statistics, the process at time $t + h$, where h is any interval, will have exactly the same statistics. Thus, for a strict stationary process,

$$x(t) - x(s) = x(t + h) - x(s + h) \qquad (6-2)$$

for any h.

Similarly, a stochastic process is stationary in the wide sense if only its mean is invariant for any change in the origin of the parameter of the process.

Let's consider one type of stochastic process that often appears in representations of real world processes. The Poisson process is defined as a counting process where the events occurring in disjoint intervals of time are independent and where the process is stationary. For a counting process, the number of occurrences of the event is nonnegative, is zero at the origin, and is monotonically increasing. For a Poisson process, the probability of any number of events occurring within any interval depends only upon the length of the interval and not where it is located

Table 6-1 Stochastic Process Type Definitions

		PARAMETER (TIME)	
		DISCRETE	CONTINUOUS
STATE SPACE	DISCRETE	DISCRETE - PARAMETER CHAIN	CONTINUOUS - PARAMETER CHAIN
	CONTINUOUS	DISCRETE - PARAMETER PRESS	CONTINUOUS - PARAMETER PROCESS

with respect to the origin. In addition, a Poisson process satisfies the following statements for a sufficiently small time interval h

P (exactly one event during h) = $\lambda h + o(h)$

P (zero events during h) = $o(h) + 1 - \lambda h$

P (more than one event during h) = $o(h)$

where $o(h)$ is some function defined by

$$\lim_{h \to \infty} \frac{o(h)}{h} = 0 \tag{6-3}$$

A function is $o(h)$ if the quantity $o(h)$ is smaller than h in the limit. For a Poisson process, the number of events occurring in a time interval of length greater than zero is represented by a random variable X, given as

$$P(X = k) = e^{-\lambda t} \frac{(\lambda t)^k}{k!} \quad k = 0, 1, 2, \ldots \tag{6-4}$$

A Poisson process is said to have an arrival rate of λ, where λ is greater than zero. The average number of arrivals (or events) during any time period then, is given as

$$\text{average number of events in } t = \lambda t \tag{6-5}$$

The previous expression for the general form of a Poisson process and the average arrival calculation are derived in [Allen 1978].

It can also be shown that if we have a Poisson process with rate λ and interarrival times $t_1, t_2 \ldots t_n$, defined by the intervals between arrivals, the interarrival times are mutually independent, identically distributed random variables, each with a mean given as

$$\mu = \frac{1}{\lambda} \tag{6-6}$$

The converse of the above can also be shown. If we have a counting process that has interarrival times that are independent, identically distributed exponential random variables with an average value defined as above, the process is a Poisson process.

The following states a useful property of the Poisson process that shows that we can find the probability that an event has occurred during a specific time interval, given that an event has occurred. Suppose that an event has occurred in the time interval from zero to t. The probability that the event occurred during a subinterval denoted s of any length is given as

$$P(\text{event occurred during } s \mid \text{an event occurred}) = \frac{t}{s} \qquad (6\text{-}7)$$

where $s \leq t, s > 0$.

Poisson processes are part of a more general family of stochastic processes called "birth and death" processes. In a birth and death process, we consider a population of size n, where the probability of an arrival, or birth, during a given interval depends on the size of the population. The probability of birth is given as

$$P(\text{birth}) = \lambda_n \, t + o(t) \qquad (6\text{-}8)$$

for an interval t. Similarly, the number of departures, or deaths, also depends on the population size. Thus, the probability of a death during an interval is given as

$$P(\text{death}) = \mu_n \, t + o(t) \qquad (6\text{-}9)$$

Formally, a birth and death process is defined as a set of states, each of which represents the populaton of the system. Equivalently, this is a continuous parameter (time) process,

$$\{X(t), t \geq 0\}$$

with each state in the state space representing a unique population. In addition, a birth and death process must satisfy the following

1. The birth and death rates are nonnegative.
2. Discrete state changes can only be made between states whose population differs by 1.
3. The probability of transition from one state to another because of a birth is given as

    $\lambda_n \, t + o(t)$

    for the time interval t.
4. The probability of transition from one state to another because of a death is given as

    $\mu_n \, t + o(t)$

for the time interval t.
5. The probability that more than one transition occurs during an interval from t to $t + d$ is given as

o(d)

and approaches zero as d approaches zero.

The above statements say that the probability of more than one birth or death within a short time interval is negligable, that no death can occur when the population equals zero, and that we can find the probability of transition from one state to an adjacent one for a given population size and time interval. We can also find, for a given time, the probability that the system has a population of n.

Suppose that we wish to find the probability that no transitions occur while in a certain state at time t (i.e., having a population of size n at time t). This can be formulated as the probability that no births occurred times the probability that no deaths occurred times the probability of having a population of size n at time t where the time interval under consideration is from t to $t + h$. This is given as follows

$$P \text{(no births)} = [1 - (\lambda_n h + o(h))] \tag{6-10}$$

$$P \text{(no deaths)} = [1 - (\mu_n h + o(h))]$$

$$P \text{(population = n at time t)} = P_n(t) \tag{6-11}$$

$$P \text{(no transitions)} = P_n(t) [1 - \lambda_n h - o(h)] [1 - \mu_n h - o(h)]$$
$$= P_n(t) [1 - \lambda_n h - o(h) - \mu_n h + \lambda_n \mu_n h^2 + \mu_n h\, o(h) - o(h) + \lambda_n h\, o(h) + o(h)^2] \tag{6-12}$$

Because $o(h)$ multiplied by a scalar quantity is still $o(h)$ and because the linear combination of any number of factors that are $o(h)$ is still $o(h)$, we obtain

$$P \text{(no transitions)} = P_n(t) [1 - \lambda_n h - \mu_n h + o(h)]$$

$$= P_n(t) [1 - \lambda_n h - \mu_n h] + o(h)] \tag{6-13}$$

Now, the probability that the system had one fewer member in the population at time t is given as

$$P \text{(population is n} - 1 \text{ at time t)} = P_{n-1}(t) \tag{6-14}$$

and the probability that a birth occurred in an interval h is given as

$$P(\text{birth in interval } t \text{ to } t + h) = \lambda_{n-1}h + o(h) \tag{6-15}$$

Thus, we get

$$P(\text{transition from population of } n - 1 \text{ to } n) = P_{n-1}(t)[\lambda_{n-1}h + o(h)]$$

$$= P_{n-1}(t)\lambda_{n-1}h + o(h) \tag{6-16}$$

again because $o(h)$ times a constant is $o(h)$.

Looking now at the probability of transitioning from a population of $n + 1$ to a population of n, we can derive a similar expression that states

$$P(\text{transition from population of } n + 1 \text{ to } n) = P_{n+1}(t)[\mu_{n+1}h + o(h)]$$

$$= P_{n+1}(t)\mu_{n+1}h + o(h) \tag{6-17}$$

We have already stated that the probability of two or more transitions during a short time interval h is negligable. Thus, we have reduced the problem of finding the probability of obtaining a population of size n during the interval t to $t + h$ to a much simpler consideration of the probabilities associated with transitioning in and out of the neighboring states (i.e., the states with population sizes of $n - 1$ and $n + 1$). We have, therefore, the probability that we will not transition out of the n element population state given that we are already in it, the probability of attaining a population of n from a population of size $n - 1$ and from a population of size $n + 1$, and the impossibility of attaining a population of n from any other state during the interval h. Since all of these events are mutually exclusive, we may sum their probabilities to find

$P(\text{population n in } t \text{ to } t + h) = P(\text{no transitions})$

$\qquad + P(\text{transition from population n} - 1 \text{ to n})$

$\qquad + P(\text{transition from population n} + 1 \text{ to n})$

$\qquad + o(h)$

$$P_n(t + h) = P_n(t)[1 - \lambda_n h - \mu_n h] + P_{n-1}(t)\lambda_{n-1}h + P_{n+1}(t)\mu_{n+1}h + o(h) \tag{6-18}$$

Rearranging and collecting terms we obtain

$$\frac{P_n(t+h) - P_n(t)}{h} = P_{n-1}(t)\lambda_{n-1} - P_n(t)(\lambda_n + \mu_n) + P_{n+1}(t)\mu_{n+1} + \frac{o(h)}{h}$$
(6-19)

The expression on the left is the definition of the derivative of $P_n(t)$ with respect to time, so letting h go to zero and taking the limit yields

$$\frac{dP_n(t)}{dt} = P_{n-1}(t)\lambda_{n-1} - P_n(t)(\lambda_n + \mu_n) + P_{n-1}(t)\mu_{n+1} \quad \text{for } n > 0$$
(6-20)

For $n = 0$, we get

$$\frac{dP_0(t)}{dt} = -\lambda_0 P_0(t) + \mu_1 P_1(t)$$
(6-21)

The initial conditions given an initial population of i are

$$P_i(0) = 1$$
(6-22)

The birth and death process, then, is defined by Equations (6-20), (6-21), and (6-22). These represent an infinite set of difference equations that theoretically has a solution for $P_n(t)$. In practice however, this solution is difficult to obtain without some simplifying assumptions. One that is frequently made is that the time-dependent property associated with a population of n is relaxed, so that

$$P_n(t) \to P_n \quad \text{as} \quad t \to \infty$$

This condition is often called the "steady state" of the system because it ignores the transitory nature of system initialization and startup. For Equations (6-20) and (6-21), if we let t approach infinity, and assume that

$$\lim_{t \to \infty} \frac{dP_n(t)}{dt} = 0$$

we obtain

$$0 = P_{n-1}\lambda_{n-1} - P_n(\lambda_n - \mu_n) + P_{n+1}\mu_{n+1} \quad n > 0$$
(6-23)

and

$$0 = \mu_1 P_1 - \lambda_0 P_0 \quad n = 0$$
(6-24)

From (6-24), we get

$$P_1 = \frac{\lambda_0 P_0}{\mu_1} \qquad (6\text{-}25)$$

If we now rearrange (6-23) as

$$P_n \mu_n - P_{n-1} \lambda_{n-1} = P_{n+1} \mu_{n+1} - P_n \lambda_n$$

and we let $n = n - 1$ in the right-hand side, we get

$$P_n \mu_n - P_{n-1} \lambda_{n-1} = P_n \mu_n - P_{n-1} \lambda_{n-1}$$

Thus, by (6-24), we get

$$P_n \mu_n - P_{n-1} \lambda_{n-1} = 0 \qquad \text{for } n > 0 \qquad (6\text{-}26)$$

We can find, therefore, a general expression for the probability of one state in terms of an adjacent state, given as

$$P_{n+1} = \frac{\lambda_n P_n}{\mu_{n+1}} \qquad (6\text{-}27)$$

Combining this result with Equation (6-25), we obtain

$$P_n = \frac{\lambda_0 \lambda_1 \ldots \lambda_{n-1}}{\mu_1 \mu_2 \ldots \mu_n} P_0 \qquad \text{for } n > 0 \qquad (6\text{-}28)$$

The initial probability of having a population of zero can be found from

$$P_0 = 1 - P_1 - P_2 - \ldots - P_\infty \qquad (6\text{-}29)$$

For the steady state probability of Equation (6-28) to exist, the server must be able to prevent infinite queue buildup. This is equivalent to stating that the probability that the system is empty must be nonzero.

Suppose that we have a server with an exponential service time distribution and arrivals from a Poisson process. If we assume that there is no waiting line at the server, so that there may be at most one customer in the system, the system will have only two states, one for a population of 0 and the other for a population of 1. By Equation (6-24) we have

$$P_1 = (\lambda/\mu) P_0 \qquad (6\text{-}30)$$

and by Equation (6-29) we know that

$$P_0 + P_1 = 1 \tag{6-31}$$

Therefore, combining Equations (6-30) and (6-31), we obtain

$$P_0 = \frac{\mu}{\lambda + \mu}$$

$$P_1 = \frac{\lambda}{\lambda + \mu}$$

If we represent each state by an oval and show the arrival and service rates into and out of each state, we obtain what is known as a "state transition diagram." Figure 6-1 shows a general state transition diagram. Because we assume that the system is in steady state, we say that the system obeys the flow balance assumption. The flow balance assumption allows us to write the state transition equations by inspection of the diagram. For example, for the diagram in Figure 6-1, we can write an expression for the probability of state 1 as

$$P_0 \lambda_0 - P_1 (\mu_1 + \lambda_1) + P_2 \mu_2 = 0 \tag{6-32}$$

In this expression, we simply pick a reference state and sum all probabilities and edges that are touching that state. By picking a consistent sign convention (e.g., all incoming edges are positive whereas outgoing edges are negative), we end up with a series of n equations plus one in the form of Equation (6-29). For the example of the single-server system with no queue that we just discussed, the state transition diagram would have only two states and the equations can be written by inspection.

MARKOV PROCESSES

A stochastic process is said to be a Markov process if the prediction of the future state of the system can be performed using only current state information without

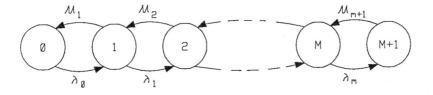

Figure 6-1 Birth and death process state transition diagram.

any knowledge of past history. More formally, if $X(t_i)$ represents the state of the system at time i, then the following is true for a Markov process

$$P(x(t_{i+1}) \mid x(t_i), x(t_{i-1}) \ldots, x(t_1)) = P(x(t_{i+1}) \mid x(t_i)) \qquad (6\text{-}33)$$

where $X(t_{i+1})$ is a future state. It can be shown that a Poisson process is also a Markov process. A Markov chain arises when the state space of the process is discrete and the process is markovian.

For a general Markov chain, the probability of transition from one state to the next generally depends upon the population of the current state. For many queuing systems, however, it is more fruitful to deal with discrete Markov processes where the transition probabilities are not state dependent. As mentioned earlier, such a process is called "stationary" or is said to have stationary transition probabilities. Another property that we will assume for our dealings with Markov chains is that every state in the process can be reached from every other state through a finite sequence of state transitions. A Markov chain with this property is called "irreducible." If we define the period of a Markov chain as the number of state transitions that occur before returning to a particular start state, the chain is said to be positive recurrent (provided it indeed does return to a given start state with average time less than infinity). In simpler terms, the chain will positively recurr if the average time of its period is finite. A Markov chain with a period of 1 is said to be aperiodic since we can build up arbitrary average times for recurrance. An ergodic Markov chain is one that satisfies the three properties just discussed for a discrete Markov chain. In summary, a Markov chain is ergodic if:

1. The Markov chain is irreducible
2. The chain is aperiodic
3. All states in the chain are positive recurrent

The fact that a Markov chain is ergodic is important for queuing theory applications because the process has a steady state probability distribution. The basic metric for determining if a Markov chain is ergodic states that if the Markov chain is aperiodic and irreducible, it is also ergodic.

A useful technique for finding the state transition probabilities involves the use of the state transition probability matrix. In it, an entry in row i and column j contains the probability of transitioning from state i to state j. Equation (6-34) shows a state transition matrix:

$$P_{ij} = \begin{vmatrix} P_{11} & P_{12} & \ldots & P_{1n} \\ P_{21} & \cdot & & \\ \cdot & & \cdot & \\ \cdot & & & \cdot \\ \cdot & & & \\ P_{n1} & \ldots & \ldots & P_{nn} \end{vmatrix} \qquad (6\text{-}34)$$

The convenience of the state transition matrix becomes apparent when calculating the probability that a transition will occur from state i to state j in n steps. It can be shown that the powers of P contain these probabilities. Thus, the second power of P shows all two transition probabilities between any two states in the chain, and so on.

QUEUING SYSTEMS

In this section, we will cover the basic analysis techniques associated with queuing systems. The prime motivation for performing queuing analysis is to assess local system behavior under a variety of assumptions, initial conditions, and operational scenarios. The modeling aspect seeks to represent the behavior of system components as processes that have calculable statistics and that adequately reflect reality. Thus, the use of queuing analysis provides us with a set of techniques for calculating quantities such as wait time for service, throughput of a server, the effect of different servers or queuing strategies, and the effects of coupled and closed networks of queues. The assumption that we must make in order to take advantage of these techniques is that the system under observation can be adequately represented by a queuing system. In the remainder of this section, we will first look at analytical modeling in general, at the characteristics of the systems that we are interested in modeling, and then at the suitability of queuing models in general and their use in particular.

What are we seeking to quantify when we set out to model a system? The answer can be summed up in just one word, performance. This one word, however, may have very different meaning for different people. Take automobile performance, for instance. For the speed enthusiast, performance is how fast the car can go and how quickly it can get to that speed. For the back road driver, it is the ability to corner without difficulty under severe conditions. For the economist, high performance means fuel efficiency and low maintenance costs. The list goes on. So it is for the performance of local area network communication environments as well. At issue here are performance measures such as the utilization of the available network bandwidth, effective throughput, average waiting time for a potential user, average number of users in the system at any given time, and the availability of service resources. In addition, trade-off analyses and "what if" studies can be performed to establish performance measures such as speedup and improved availability — in general, the sensitivity of the previously mentioned measures to changes in the system under study.

The general process of analytical modeling involves mapping the behavior of a complex system onto a relatively simpler system, solving the simpler system for the measures of interest, and then extrapolating the results back to the complex system. Sometimes this process has several levels where models are broken into submodels. Here, the lowest-level models are solved (or partially solved) first, their results propagated up to the next higher layer for inclusion in that layer's solution, and so on to the top level.

In some cases, portions of a model can be replaced by a technique called "decomposition." Here, a queuing subsystem is replaced with a flow-equivalent server where the server output is precalculated for each number of units (or customers) in the system. Thus, the job flow through the flow-equivalent server can be implemented using a simple lookup table indexed by the number of customers currently in the system. This technique is appropriate if the impact of the removed subsystem is minimal when compared to the effect of other model subsystems. See [Allen 1978] for a discussion of flow-equivalent servers.

The basic premise behind the use of queuing models for computer systems analysis is that the components of a computer system can be represented by a network of servers (or resources) and waiting lines (queues). A server is defined as an entity that can affect, or even stop, the flow of jobs through the system. In a computer system, a server may be the CPU, an I/O channel, memory, or a communication port. A waiting line is just that, a place where jobs queue for service. To make a queuing model work, jobs (or customers or message packets or anything else that requires the sort of processing provided by the server) are inserted into the network. A simple example, the single server model, is shown in Figure 6-2. In that system, jobs arrive at some rate, queue for service on a first-come-first-served basis, receive service, and exit the system. This kind of model, with jobs entering and leaving the system, is called an "open" model.

By cascading simple queuing models and allowing the existence of parallel servers, networks of queues and servers may be formed. These combinations are formally called "queuing networks," although we will also call them "network models" and "queuing systems." Figure 6-3 shows one such model of a computer system with a fixed number of jobs competing for two CPUs and an I/O processor.

In Figure 6-3, jobs that have finished I/O service loop back into the CPU queue for another cycle of computation and I/O. A system like this, where the number of customers remains constant, is called a "closed" system.

A combination of the open and closed concepts is certainly possible if one considers each job to have an associated class. For example, a computer system may contain two job classes, interactive and system, where interactive jobs come and go as users log on and off and where system jobs execute continually. A system that contains both open and closed class customers is called "mixed."

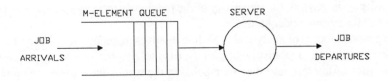

Figure 6-2 A single-server queuing model.

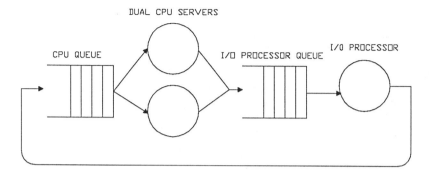

Figure 6-3   A queuing network.

The concept of customer classes also allows different classes to receive different treatment at the same server, as well as the definition of a group of customers as open or closed. A system with more than one customer class is called "multiclass," and it may be either open, closed, or mixed.

Once we have a network model established, the collection of n_1 customers at server 1, n_2 at server 2, and so on for the entire system, defines the state of the network model. An analytical model for a queuing network would provide a method for calculating the probability that the network is in a particular state. In addition, network throughput, mean queue length for any server, and mean response time (wait time and service time) for any server can be found by a variety of methods.

In a network model, a server typically has associated with it a service time distribution from which customer service times are drawn. Upon arrival at a server, a customer receives service, the duration of which is determined by the service time distribution.

We will now turn our attention to some of the more well-known queuing systems, the notation used to represent them, the performance quantities of interest, and the methods for calculating them. We have already introduced many notations for the quantities of interest for random variables and stochastic processes. Figure 6-4 reviews these and adds a host of others that will be useful for the analysis of queuing systems. The paragraphs that follow briefly discuss the more important parameters.

The arrival rate for a queuing system defines the stream of arrivals into a queue from some outside source. This rate is defined as an average rate that is derived from an arrival process that may well be one of those discussed earlier. The average interarrival time for a given arrival process is denoted as

$$E[\tau] = \frac{1}{\lambda} \qquad (6\text{-}35)$$

λ - ARRIVAL RATE AT ENTRANCE TO A QUEUE
μ - SERVICE RATE (AVERAGE) OF A SERVER
P_n - PROBABILITY THAT THERE ARE n CUSTOMERS IN THE SYSTEM AT STEADY STATE
C - NUMBER OF IDENTICAL SERVERS IN THE QUEUING SYSTEM
N - RANDOM VARIABLE FOR THE NUMBER OF CUSTOMERS IN THE SYSTEM AT STEADY STATE
L - E[N], EXPECTED NUMBER OF CUSTOMERS IN THE SYSTEM AT STEADY STATE
W_q - RANDOM VARIABLE FOR CUSTOMER WAITING TIME IN A QUEUE
S - RANDOM VARIABLE FOR CUSTOMER SERVICE TIME
N_q - RANDOM VARIABLE FOR THE NUMBER OF CUSTOMERS IN A QUEUE AT STEADY STATE
L_q - E[N_q], EXPECTED NUMBER OF CUSTOMERS IN A QUEUE AT STEADY STATE
N_s - RANDOM VARIABLES FOR THE NUMBER OF CUSTOMERS AT A SERVER AT STEADY STATE
W - W_q + S, RANDOM VARIABLE FOR THE TOTAL WAITING TIME IN A SYSTEM

Figure 6-4 Notation for queuing systems.

The service rate parameter is defined in a way that is similar to the arrival rate. This rate is also an average rate that defines how many customers are processed per unit time when the server is busy. The service rate can be cast in terms of the service time random variable as

$$\mu = \frac{1}{E[s]} \qquad (6\text{-}36)$$

Often, we wish to know the probability that the system will contain exactly n customers at steady state. Accounting for all of the probabilities for n ranging from zero to infinity defines the probability distribution for the number of customers in the system.

The number of identical servers in a system indicate that a customer leaving a queue may proceed to one of C servers as soon as one becomes nonbusy.

Of interest for any queuing system is the average number of customers in the system at steady state. This value can be thought of as the sum of all customers in queues and at servers,

$$N = N_q + N_s$$

$$L = E[N] = E[N_q] + E[N_s] \qquad (6\text{-}37)$$

The total time a customer spends in the system can also be thought of as the sum of wait time in the queues and time at the servers. The wait time and expected wait time at steady state, therefore, are given as

```
A/B/c/K/m/Z
```

WHERE

```
A - ARRIVAL PROCESS DEFINITION
B - SERVICE TIME DISTRIBUTION
c - NUMBER OF IDENTICAL SERVERS
K - MAXIMUM NUMBER OF CUSTOMERS ALLOWED IN THE SYSTEM (DEFAULT = ∞)
m - NUMBER OF CUSTOMERS ALLOWED TO ARRIVE BEFORE THE ARRIVAL PROCESS
    STOPS (DEFAULT = ∞)
Z - DISIPLINE USED TO ORDER CUSTOMERS IN THE QUEUE (DEFAULT = FIFO)
```

Figure 6-5 The Kendall notation.

$w = q + s$

$$E[w] = E[q] + E[s] \tag{6-38}$$

In addition to the notation described above for the quantities associated with queuing systems, it is also useful to introduce a notation for the parameters of a queuing system. The notation we will use here is known as the "Kendall" notation, illustrated in Figure 6-5.

The symbols used in a Kendall notation description also have some standard definitions. Figure 6-6 shows the more common designators for the A and B fields of the notation.

The service discipline used to order customers in the queue can be any of a variety of types such as first-in-first-out (FIFO), last-in-first-out (LIFO), priority ordered, random ordered, and others. Next, we will examine several queuing systems and give expressions for the more important performance quantities.

The $M/M/1$ System

The $M/M/1$ queuing system is characterized by a Poisson arrival process and exponential service time distributions, with one server, and a FIFO queue ordering discipline. The system, shown in Figure 6-7, may represent an input buffer holding incoming data bytes, with an I/O processor as the server. In such an application, it

```
D - DETERMINISTIC SERVICE TIME OR ARRIVAL RATE
G - GENERAL SERVICE TIME OR ARRIVAL RATE
M - MARKOVIAN (EXPONENTIAL) SERVICE TIME OR ARRIVAL RATE
```

Figure 6-6 Common service time and arrival rate designators.

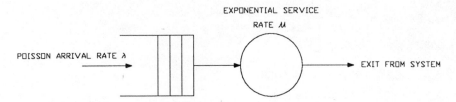

Figure 6-7 *M/M/1* queuing system.

would be useful to analyze the queue size and average service time for a byte at the input channel.

A few of the quantities that we will be interested in for this type of queuing system are the average queue length, the wait time for a customer in the queue, the total time a customer spends in the system, and the server utilization. Let's look at the exponential service distribution first. It is given as

$$s = \mu e^{-\mu t}$$

and is shown in Figure 6-8. In the figure, $E[s]$ is the average service time of a customer at the server. Next, let's derive the steady state equations for the *M/M/1* system.

The *M/M/1* system is a birth and death process as discussed earlier. Let us assume that

$P_n(t)$ = probability that n customers are in the system at time t

From earlier discussions about birth and death processes, we know that [see Equation (6-18)]

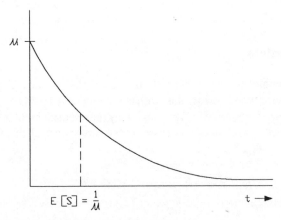

Figure 6-8 Exponential service time distribution.

$$P_n(t+h) = P_n(t)[1 - \lambda_n h - \mu_n h] + P_{n-1}(t)\lambda_{n-1} h + P_{n+1}(t)\mu_{n+1} h + o(h)$$

and that

$$P_0(t+h) = P_0(t) - P_0(t)\lambda h + P_1(t)\mu h + o(h)^2$$

Following the same reasoning for deriving the steady state probabilities as we did for the general birth and death process, we obtain the steady state equations for the *M/M/*1 system.

$$\lambda P_0 = \mu P_1 \tag{6-39}$$

$$(\lambda + \mu) P_n = \lambda P_{n-1} + \mu P_{n+1} \quad \text{for } n > 0 \tag{6-40}$$

Now, if we let u denote the average server utilization, we define this quantity as the mean service time of a single customer divided by the mean interarrival time [see Equations (6-35) and (6-36)], then

$$u = \frac{1/\mu}{1/\lambda} = \frac{\lambda}{\mu} \tag{6-41}$$

Solving the steady state Equations (6-39) and (6-40), we obtain

$$P_1 = \frac{\lambda}{\mu} P_0 \qquad \text{for } n = 0$$

$$P_2 = (\lambda + \mu) P_1 - \lambda P_0$$

$$= (1 + \frac{\lambda}{\mu}) P_1 - \frac{\lambda}{\mu} P_0$$

$$= (1 + \frac{\lambda}{\mu}) \frac{\lambda}{\mu} P_0 - \frac{\lambda}{\mu} P_0$$

$$P_2 = \left(\frac{\lambda}{\mu}\right)^2 P_0 \qquad \text{for } n = 1$$

Similarly,

$$P_3 = \left(\frac{\lambda}{\mu}\right)^3 P_0$$

$$P_n = \left(\frac{\lambda}{\mu}\right)^n P_0 \quad \text{for } n > 0$$

$$P_n = u^n P_0 \quad \text{for } n > 0 \tag{6-42}$$

We assume here that u is less than 1 so that we have a finite queue length. Now, we know that

$$\sum_{n=0}^{\infty} P_n = 1$$

and that

$$\sum_{n=0}^{\infty} P_n = P_0 \sum_{n=0}^{\infty} u^n = 1$$

so

$$P_0 = \frac{1}{\sum_{n=0}^{\infty} u^n} \tag{6-43}$$

The right-hand side of Equation (6-43) is recognized as a geometric progression that has the following solution:

$$P_0 = \frac{1}{1/(1-u)}$$

$$P_0 = 1 - u = 1 - \frac{\lambda}{\mu} \tag{6-44}$$

Combining Equations (6-42) and (6-44), we arrive at the steady state probability that there are n customers in an $M/M/1$ system.

$$P_n = (1 - \frac{\lambda}{\mu})\left(\frac{\lambda}{\mu}\right)^n \tag{6-45}$$

Figure 6-9 shows the state transition diagram for the *M/M/*1 queuing system.
Now let's look at the average number of customers in the system at steady state. This is given as the expected value of *N*, which can be found by

$$E[N] = \sum_{n=0}^{\infty} nP_n$$

$$= \sum_{n=0}^{\infty} n(1 - \frac{\lambda}{\mu})\left(\frac{\lambda}{\mu}\right)^n$$

$$= (1 - \frac{\lambda}{\mu})\sum_{n=0}^{\infty} n\left(\frac{\lambda}{\mu}\right)^n$$

$$= (1 - \frac{\lambda}{\mu})\left(\frac{\lambda}{\mu} + 2\left(\frac{\lambda}{\mu}\right)^2 + 3\left(\frac{\lambda}{\mu}\right)^3 + ...\right)$$

$$= (1 - \frac{\lambda}{\mu})\frac{\lambda}{\mu}\left(1 + 2\frac{\lambda}{\mu} + 3\left(\frac{\lambda}{\mu}\right)^2 + ...\right)$$

$$= (1 - \frac{\lambda}{\mu})\frac{\lambda}{\mu}\sum_{n=1}^{\infty} n\left(\frac{\lambda}{\mu}\right)^{n-1}$$

$$= \frac{(1 - \frac{\lambda}{\mu})\frac{\lambda}{\mu}}{(1 - \frac{\lambda}{\mu})^2}$$

$$E[N] = \frac{\frac{\lambda}{\mu}}{1 - \frac{\lambda}{\mu}} \tag{6-46}$$

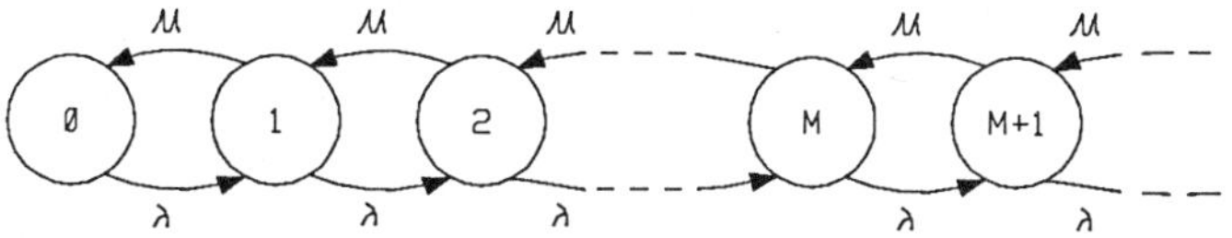

Figure 6-9 *M/M/*1 system state transition diagram.

The average amount of time that a customer must wait in the queue, assuming that other customers are already in the queue, is given as the number of customers ahead divided by the average service time of the server.

$$E[w_q | n = i] = \frac{i}{\mu} \tag{6-47}$$

The expected wait time in the queue, then, is a function of the average wait time and the steady state probability of having i customers in the system.

$$E[w_q] = \sum_{i=1}^{\infty} \frac{i}{\mu} P_i \tag{6-48}$$

$$= \frac{1}{\mu} E[N]$$

$$E[w_q] = \frac{\frac{\lambda}{\mu^2}}{1 - \frac{\lambda}{\mu}} \tag{6-49}$$

Combining the queue waiting time [Equation (6-49)] and the expected service time E[s] [Equation (6-36)] yields the total customer time in the system, called the "expected wait time."

$$E[w] = E[w_q] + E[s] \tag{6-50}$$

$$= \frac{\frac{\lambda}{\mu^2}}{1 - \frac{\lambda}{\mu}} + \frac{1}{\mu}$$

$$= \frac{1}{\mu} \left(\frac{\frac{\lambda}{\mu}}{1 - \frac{\lambda}{\mu}} + 1 \right)$$

$$= \frac{1}{\mu} \left(\frac{1}{1 - \frac{\lambda}{\mu}} \right)$$

$$E[w] = \frac{1}{\mu - \lambda} \tag{6-51}$$

If we rewrite Equation (6-46) as

$$E[N] = \frac{\lambda}{\mu - \lambda}$$

and compare it with Equation (6-51), we can obtain what is known as "Little's result":

$$E[N] = \lambda E[w] \tag{6-52}$$

Little's result holds in general for any queuing system in steady state that conforms to the flow balance assumption discussed earlier. As such, it gives us an important relationship for the effect of arrival rate and queue length on total customer wait time. A related result, also attributed to Little, states the equivalent for queue length and queue waiting time and also holds for queuing systems in steady state:

$$E[N_q] = \lambda E[w_q] \tag{6-53}$$

This second version of Little's result says that the expected queue length can be found directly from the arrival rate times the expected queue wait time.

The total waiting time in the system, then, can be found by using Little's result or by summing the queue wait time and the expected service time.

Server utilization is a useful quantity for determining how many equivalent servers must be provided to service a given arrival process. The method is staightforward and involves solving an $M/M/1$ queuing system using the methods indicated above. Suppose, for instance, that we have an $M/M/1$ system with an arrival rate of six customers per minute and a service time of 10 seconds. Then the server utilization, as given by Equation (6-41), is 1. This means that the server can be expected to be busy 100 percent of the time but that it can, in fact, process enough customers so that infinite queue buildup is prevented. Suppose now that the arrival rate increases to 60 customers per minute so that the server utilization becomes 10, an overload situation. If we speed up the server by 10, however, or provide 10 servers of the original speed, the utilization would again be 1. In general, then, if the utilization is less than 1, the server can keep up with the flow of customers and an infinite queue will not result. If, however, the utilization is greater than 1, the utilization, rounded up to the next largest integer, gives an indication of the number of identical servers that are necessary to keep up with the customer flow.

A final interesting property of the $M/M/1$ queuing system is the fact that the queue waiting time and total waiting time both have exponential distributions. For instance, the queue wait time can be found as follows:

$$P[0 < w_q \le t] = \sum_{N=1}^{\infty} P_N \int_0^t f_{w_q|N}(X|N)\, dx$$

From Equation (6-42) and the distribution (Poisson), we get

$$P[0 < w_q \le t] = \left(\frac{\lambda}{\mu}\right)^N \left(1 - \frac{\lambda}{\mu}\right) \int_0^t \frac{\mu^N x^{N-1}}{(N-1)!} e^{-\mu x}\, dx$$

$$= \int_0^t \lambda e^{-\mu x} \left(1 - \frac{\lambda}{\mu}\right) \sum_{N=1}^{\infty} \frac{(\lambda x)^{N-1}}{(n-1)!}\, dx$$

$$= \int_0^t \lambda e^{-\mu x} \left(1 - \frac{\lambda}{\mu}\right) e^{\lambda x}\, dx$$

$$= \frac{\lambda}{\mu} \int_0^t (\mu - \lambda) e^{-x(\mu - \lambda)}\, dx$$

$$= \frac{\lambda}{\mu} [1 - e^{-t(\mu - \lambda)}]$$

From Equation (6-51), we substitute to get

$$P[0 < w_q \le t] = \frac{\lambda}{\mu} [1 - e^{-t/E[w]}]$$

Hence,

$$P[w_q \le t] = P_0 + \frac{\lambda}{\mu} [1 - e^{-t/E[w]}]$$

$$P[w_q \le t] = 1 - \frac{\lambda}{\mu} e^{-t/E[w]} \tag{6-54}$$

Similarly,

$$P[w_q \le t] = 1 - e^{-t/E[w]}$$

From these distributions, we can find the percentiles for the expected wait time for r percent of the total number of customers. The percentile of any random variable is defined as

$$P[x \leq \pi(r)] = \frac{r}{100} \tag{6-55}$$

In the case of queue wait time, for example, if we wish to find the wait time that 90 percent of the customers in the system will not exceed, we have

$$1 - e^{-\pi(90)/E[w]} = 0.9$$

$$0.1 = e^{-\pi(90)/E[w]}$$

$$\ln(0.1) = -\pi(90)/E[w]$$

$$\pi(90) = 2.3\, E[w] \tag{6-56}$$

The $M/M/1/K$ System

An interesting and realistic variation on the basic $M/M/1$ system is a system with a finite queue size. In this system, once the queue is full, new arrivals are lost and are never provided service. This is quite realistic, for example, in an input system with finite input buffer space and no flow control protocol. The birth-death state transition diagram for the $M/M/1/K$ system is shown in Figure 6-10.

As for the $M/M/1$ system, we have

$$P_n = \left(\frac{\lambda}{\mu}\right)^n P_0 \qquad \text{for } K \geq n \geq 0 \tag{6-57}$$

Using the law of total probability, we also have

$$\sum_{i=0}^{K} P_i = 1$$

$$\sum_{i=0}^{K} \left(\frac{\lambda}{\mu}\right)^i P_0 = 1$$

$$P_0 \sum_{i=1}^{K} \left(\frac{\lambda}{\mu}\right)^i = 1$$

140 Modeling and Analysis of Local Area Networks

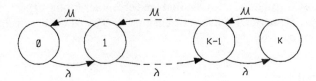

Figure 6-10 State transition diagram for the $M/M/1/K$ system.

The summation is a geometric series, which yields

$$P_0 \frac{(1-\left(\frac{\lambda}{\mu}\right)^{K+1})}{1-\frac{\lambda}{\mu}} = 1 \quad \text{if } \lambda \neq \mu$$

so

$$P_0 \frac{1-\frac{\lambda}{\mu}}{1-\left(\frac{\lambda}{\mu}\right)^{K+1}} \tag{6-58}$$

Substituting into Equation (6-57) yields

$$P_n = \frac{1-\frac{\lambda}{\mu}}{1-\left(\frac{\lambda}{\mu}\right)^{K+1}} \left(\frac{\lambda}{\mu}\right)^n \quad \text{for } K \geq n \geq 0 \tag{6-59}$$

If the arrival rate is equal to the service rate, we have

$$P_0 \sum_{i=0}^{K} \left(\frac{\lambda}{\mu}\right)^i = 1 \quad \text{for } \lambda = \mu$$

$$P_0 = \frac{1}{K+1} \quad \text{for } \lambda = \mu \tag{6-60}$$

and

$$P_n = \frac{1}{K+1} \quad \text{for } K \geq n \geq 0 \quad \text{and} \quad \lambda = \mu \tag{6-61}$$

The expected number of customers in the system, for a system with nonequal arrival and service rates, is found as

$$E[N] = \sum_{i=0}^{K} i P_i$$

$$= \sum_{i=0}^{K} i \left[\frac{1-\frac{\lambda}{\mu}}{1-\left(\frac{\lambda}{\mu}\right)^{K+1}}\right] \left(\frac{\lambda}{\mu}\right)^i$$

$$E[N] = \frac{1-\frac{\lambda}{\mu}}{1-\left(\frac{\lambda}{\mu}\right)^{K+1}} \sum_{i=0}^{K} i \left(\frac{\lambda}{\mu}\right)^i$$

After some algebra and simplification of the summation, we get

$$E[N] = \frac{\frac{\lambda}{\mu}}{1-\frac{\lambda}{\mu}} - \frac{(K+1)\left(\frac{\lambda}{\mu}\right)^{K+1}}{1-\left(\frac{\lambda}{\mu}\right)^{K+1}} \quad (6\text{-}62)$$

We can see that, for very large values of K, the second term approximates zero and we get the same expression as for the $M/M/1$ system. For the case where the arrival and service rates are equal,

$$E[N] = \sum_{i=0}^{K} i P_i$$

$$E[N] = \frac{1}{K+1} \sum_{i=1}^{K} n$$

$$E[N] = \frac{K}{2} \quad \text{for } \lambda = \mu \quad (6\text{-}63)$$

To compute the wait time distribution for the $M/M/1/K$ system, we must compute the probability for the number of customers in the queue when a customer arrives, given that the customer is admitted to the system. This is given as

$$P(\text{n customers in system} \mid N < K) = \frac{P_N}{1 - P_K} \qquad (6\text{-}64)$$

From this, we can arrive at the wait time distribution

$$P(w \le t) = 1 - \sum_{N=0}^{K-1} \frac{P_N}{1-P_K} \sum_{K=0}^{N} e^{-\mu t} \frac{(\mu t)^K}{K!} \qquad (6\text{-}65)$$

This quantity can be found in the same way as the statistic for the *M/M/1* system.

The *M/M/C* System

The *M/M/C* system, shown in Figure 6-11, consists of a single waiting line that feeds C identical servers. The arrival process is considered to be Poisson, and the servers have exponential service times. The state transition diagram for an *M/M/C* system is shown in Figure 6-12. Now let's write some of the flow balance equations for the state transition diagram.

$$P_1 = \frac{\lambda}{\mu} P_0$$

$$P_2 = \frac{\lambda^2}{2\mu^2} P_0$$

$$P_3 = \frac{\lambda^3}{6\mu^3} P_0$$

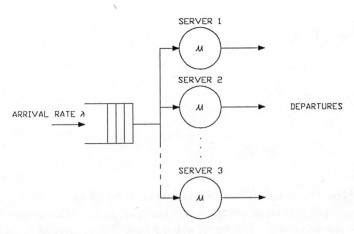

Figure 6-11 An *M/M/C* system.

Figure 6-12 State transition diagram for an *M/M/C* system.

$$P_C = \frac{\lambda^C}{C!\mu^C} P_0$$

$$P_{C+1} = \frac{\lambda^{C+1}}{CC!\mu^{C+1}} P_0$$

$$\vdots$$

$$P_N = \frac{\lambda^N}{C^{N-C}C!\mu^N} P_0$$

so that

$$P_N = \begin{cases} \dfrac{\lambda^N}{N!\mu^N} P_0 & \text{for } N \leq C \\ \dfrac{\lambda^N}{C^{N-C} C!\mu^N} & \text{for } N > C \end{cases} \tag{6-66}$$

To find the probability of no customers in the system, we sum all probabilities.

$$1 = P_0 \left[1 + \frac{\lambda}{\mu} + \left(\frac{\lambda}{\mu}\right)^2 \Big/ 2 + \ldots \left(\frac{\lambda}{\mu}\right)^C \Big/ C! + \left(\frac{\lambda}{\mu}\right)^{C+1} \Big/ CC! + \ldots + \left(\frac{\lambda}{\mu}\right)^N \Big/ C^{N-C} C! + \ldots \right]$$

$$1 = P_0 \sum_{i=1}^{C} \left(\frac{\lambda}{\mu}\right)^i \Big/ i! + \left(\frac{\lambda}{\mu}\right)^C \Big/ C! \, (1 - \frac{\lambda}{c\mu}) \tag{6-67}$$

The expected queue length can be found by subtracting the number of customers in service from the expected number of customers in the system:

$$E[N_q] = \sum_{N=C}^{\infty} (N-C) P_N$$

$$E[N_q] = \frac{P_0 \left(\frac{\lambda}{\mu}\right)^C \left(\frac{\lambda}{C\mu}\right)}{C!\,(1-\frac{\lambda}{C\mu})^2} \tag{6-68}$$

Using Little's result, we can compute the queue wait time:

$$w_q = \frac{E[N_q]}{\lambda} \tag{6-69}$$

the total wait time:

$$w = w_q + E[s]$$

$$w = \frac{E[N_q]}{\lambda} + \frac{1}{\mu} \tag{6-70}$$

and the total number of customers in the system:

$$E[N] = \lambda w$$

$$E[N] = E[N_q] + \frac{\lambda}{\mu} \tag{6-71}$$

For some multiple server systems, no queue is provided for customers to wait for service. In this case, a customer that arrives when all servers are busy is turned away, perhaps to try again later. The state transition diagram is shown in Figure 6-13. This system is often referred to as the *M/M/C* loss system because customers that arrive when all servers are busy are lost. Writing the flow balance equations, we obtain the steady state probabilities as we did for the *M/M/C* system:

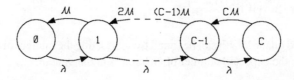

Figure 6-13 *M/M/C* loss system.

$$P_N = \frac{\left(\frac{\lambda}{\mu}\right)^N / N!}{1 + \frac{\lambda}{\mu} + \left(\frac{\lambda}{\mu}\right)^2 / 2! + \ldots \left(\frac{\lambda}{\mu}\right)^C / C!} \tag{6-72}$$

The probability that a customer will be turned away, then, can be found from the above expression with $N = C$. Since there is no queue, the queue length and queue waiting time are zero, and the total wait time is the expected service time.

The M/G/1 System

The queuing systems that we have discussed so far have all had the Markov property for arrival and service processes, making it possible to model the system as a birth and death process and to write the flow balance equations by inspection. Next, we will look at a system in which the service time does not have the Markov property. In the M/G/1 system, each customer has different and independent service times. Because service times are not guaranteed to be markovian, the system is not representative of a Markov chain and we must resort to other methods to derive meaningful statistics. One approach that is commonly taken is to look at the process that describes jobs leaving the system, which is itself a stochastic process that also happens to be a Markov chain. It has been shown [Kleinrock 1975] that, in the limit, the distribution for the number of jobs in the system at any point in time and the number of jobs in the system observed precisely when a customer departs from the system, are identical. Summarizing the procedure, then, we can analyze certain aspects of the system that are described by a nonmarkovian process by observing a markovian subportion of the system (in this case the departure process) and extrapolating the results back to the original system. This type of analysis relies on what is known as an "embedded" Markov chain. The derivation of the statistics for the M/G/1 system is beyond the scope of this book, however; the following results are derived in [Allen 1978] or [Trivedi 1982].

Because we can relate the completion of service to an embedded Markov chain, we still claim the following for utilization:

$$u = \lambda E[s] \tag{6-73}$$

The expected number of customers in the queue is given as

$$E[N_q] = \frac{\lambda E[s^2]}{2(1 - \frac{\lambda}{\mu})} \tag{6-74}$$

Using Little's result, we get the average time spent waiting in the queue:

$$w_q = \frac{E[N_q]}{\lambda} = \frac{E[s^2]}{2(1-\frac{\lambda}{\mu})} \qquad (6\text{-}75)$$

The total number of customers in the system, on average, is given as

$$E[N] = E[N_q] + \frac{\lambda}{\mu} = \frac{\lambda E[s^2]}{2(1-\frac{\lambda}{\mu})} + \frac{\lambda}{\mu} \qquad (6\text{-}76)$$

and the expected customer wait time, again using Little's result, is

$$w = \frac{E[N]}{\lambda} = \frac{E[s^2]}{2(1-\frac{\lambda}{\mu})} + \frac{1}{\mu} \qquad (6\text{-}77)$$

The general $M/G/1$ system is useful in many situations because we can characterize a known service process in terms of its moments and then evaluate its performance in the presence of a "random" arrival process.

The $G/M/1$ System

In the previous section, we discussed the situation in which a system had non-markovian service process. Next, we will consider the case in which the service time is random and the arrival process is nonmarkovian. We will assume that the interarrival times are independent and identically distributed. Again, we can find an embedded Markov chain in this system whose behavior is essentially equivalent to the system's behavior at steady state. In this case, the random variable defining the number of customers in the system, at precisely the time when another arrival occurs, forms a process that is a Markov chain. As with the other systems that we have discussed, the statistics of interest use the probability of having an empty system in calculating their values. It has been shown, for a general arrival process, that the probability that an arrival finds the system empty can be found by solving the following equation:

$$1 = P[N=0] + E[e^{-\mu^{P[N=0]}}] \qquad (6\text{-}78)$$

The second term on the right-hand side of this equation is the Laplace transform of the arrival density function $f(x)$ with

$$\theta = -\mu P[N=0]$$

and is defined as

$$E = [e^{-\mu P_0}] = \begin{cases} \int_0^\infty e^{-\mu P_0 x} f(x)dx & \text{for a continuous process} \\ \sum_{i=0}^\infty e^{-\mu P_0 x_i} f(x) & \text{for a continuous process} \end{cases} \quad (6\text{-}79)$$

Combining Equations (6-78) and (6-79) gives us an expression with which we can solve for the probability of having an empty system. Once we have found this quantity, we can use it in the following equations to find the expected number of customers in the system at steady state, the expected wait time for a customer, the expected queue length, and the expected queuing time:

$$E[N] = \frac{\lambda}{\mu P_0} \quad (6\text{-}80)$$

$$w = \frac{1}{\mu P_0} \quad (6\text{-}81)$$

$$E[N_q] = \frac{\lambda}{\mu}\left(\frac{1-P_0}{P_0}\right) \quad (6\text{-}82)$$

and $\quad w_q = \dfrac{1-P_0}{\mu P_0}$

NETWORKS OF QUEUES

Until now, we have been considering queuing systems that contain only one "station." That is, the systems that we have looked at have a single queue and a single server or set of servers, and customers arrive only at that queue and depart only following service. This situation is fine for relatively simple systems that are either not connected to other systems or that can be considered isolated from other, connected systems. Now, we will consider the case in which several queuing systems are interconnected and attempt to find meaningful statistics on such a system's behavior.

Referring to the beginning of the queuing systems section, we recall that a network of queues results from connecting the departure stream of one queuing system to the queue input of another, for an arbitrary number of queuing systems connected in an arbitrary way. Also, we discussed the concept of open and closed networks in which an open network was defined as one in which arrivals from, and

148   Modeling and Analysis of Local Area Networks

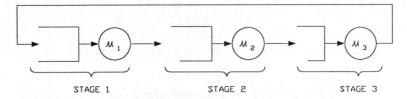

Figure 6-14   A three-stage closed queuing network.

departures to, the outside world are permitted and a closed network as one in which they are not permitted. We will discuss general classes of both types here.

Closed Networks

Consider the closed three-stage network of queues in Figure 6-14. Assume that the service time for each server is exponentially distributed and unique to that server and that the system contains two customers. We can describe this system as a Markov process with each state in the process defined as the triplet

$$\text{state} = \{N_1, N_2, N_3\} \tag{6-83}$$

where N_i is the number of customers in the i queue. Also, since we have two customers,

$$\sum_{i=1}^{3} N_i = 2$$

The state transition diagram for the system, with the states labeled as defined in Equation (6-83), is shown in Figure 6-15. The labels on the edges denote customer movements from stage to stage and are dependent upon the service rate for the stage from which a customer is departing.

To find the steady state probabilities for each state in the system, we can write what are known as the "flow balance equations." As discussed earlier, the flow balance assumption states that we can represent the steady state probabilities of a Markov process by writing the equations that balance the flow of customers into and out of the states in the network. For each individual state, then, we can write a balance equation that equates flow into a state with flow out of a state. For the states of Figure 6-15, then, we can write the following balance equations:

$$\pi(2, 0, 0)\, \mu_1 = \pi(1, 0, 1)\, \mu_3$$

$$\pi(1, 1, 0)\, (\mu_1 + \mu_2) = \pi(2, 0, 0)\, \mu_1 - \pi(0, 1, 1)\, \mu_3$$

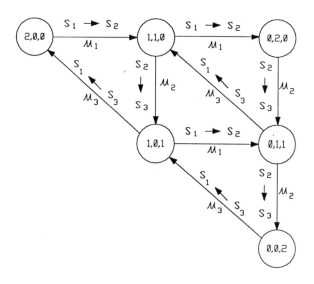

Figure 6-15 State transition diagram for a simple closed system.

$\pi(0, 2, 0)\mu_2 = \pi(1, 1, 0)\mu_1$

$\pi(1, 0, 1)(\mu_3 + \mu_1) = \pi(1, 1, 0)\mu_2 + \pi(0, 0, 2)\mu_3$

$\pi(0, 1, 1)(\mu_2 + \mu_3) = \pi(0, 2, 0)\mu_2 + \pi(1, 0, 1)\mu_1$

$\pi(0, 0, 2)\mu_3 = \pi(0, 1, 1)\mu_2$

when $\pi(N_1, N_2, N_3)$ = Probability of state $\{N_1, N_2, N_3\}$

Keeping in mind that the sum of all of the state probabilities must equal 1, this network has the solution

$$\pi(N_1, N_2, N_3) = K \left(\frac{1}{\mu_1}\right)^{N_1} \left(\frac{1}{\mu_2}\right)^{N_2} \left(\frac{1}{\mu_3}\right)^{N_3}$$

where K is a normalization constant to ensure that the probabilities sum to 1.

$$K = \frac{1}{\sum_{i=0}^{2}\sum_{j=0}^{2}\sum_{k=0}^{2} \pi(i, j, k)} \tag{6-84}$$

150 Modeling and Analysis of Local Area Networks

Now, we can solve the flow balance equations for the individual state probabilities. Once we have the state probabilities, we can find the expected length at any of the servers as follows:

$$E[N_q] = \sum_{i=1}^{\text{\# states}} i\, P[N=i \text{ at queue k}] \tag{6-85}$$

Because the state probabilities at the queues are not the same as for the queuing systems in isolation, we cannot find the expected wait time in a queue by simply multiplying the number of customers by the service time at that queue. Instead, we shall first calculate the throughput for each queuing system and then apply Little's result using the throughput as a measure of the arrival rate at a particular queue. Therefore, the throughput at a particular queue can be found by multiplying the probability of having a customer in that queue (e.g., the server is busy) times the expected service rate:

$$\lambda_i = P(\text{server is busy})\, \mu \tag{6-86}$$

Now, using Little's result, we can calculate the time spent in each queue by a customer at the respective queues:

$$w_q = E[N_k]/\lambda_i \tag{6-87}$$

The total round trip waiting time for a customer in the system can be found by summing all of the queue waiting times. It can also be found directly by using Little's result and the average throughput for the system. Thus, for two customers, we have

$$w_q = \frac{2}{\lambda_{ave}} \tag{6-88}$$

Next, let's consider an arbitrary closed network with M queues and N customers. Assume that all servers have exponential service time distributions. For the sake of discussion, the network in Figure 6-16 will represent our arbitrary network. Let's define a branching probability as the probability of having any customer follow a particular branch when arriving at a branch point. Therefore, let

P_{ij} = Probability that a customer leaving server i goes to queue j

For any server i,

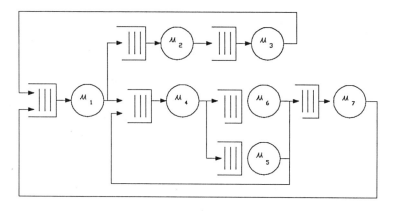

Figure 6-16 An arbitrary network of queues and servers.

$$\sum_{\text{all } j} P_{ij} = 1$$

The conservation of flow in the system requires that

$$\lambda_j = \sum_{i=1}^{M} \lambda_i P_{ij} \qquad (6\text{-}89)$$

Define the relative throughput of a server i as

$$B(j) = \sum_{i=1}^{M} B(i) P_{ij} \qquad (6\text{-}90)$$

Since the B terms are relative, we can arbitrarily set one of them equal to 1 and solve for the others. Once we have all of the terms, the steady state probabilities are given by

$$P(N_1, N_2, N_3, \ldots N_M) = K \prod_{i=1}^{M} \left(\frac{B(i)}{\mu_i}\right)^{N_i} \qquad (6\text{-}91)$$

Equation (6-90) can be derived by assuming the conservation of flow for a particular state and then by solving the system of equations as we did for the previous example. Let's pick a state S so that

$$S = (k_1, k_2, \ldots k_M)$$

152 Modeling and Analysis of Local Area Networks

and examine the effects of arrivals and departures of customers from queue j. Define another state A that is identical to S except that it has one more customer at queue i and one less customer at queue j than S. Thus, A is a neighbor state of S. We are postulating that the rate of entering state S due to an arrival at queue j is balanced by the rate of leaving state S due to a departure from queue j. Since there may be more than one state A where there is one more customer at queue i and one less at j, we must balance all such states against state S. Equating the flows then, results in

$$\sum_{i=1}^{M} P[A_i] \mu_i P_{ij} = P[s] \mu_j \qquad (6\text{-}92)$$

from (6-91),

$$P[A_i] = K \prod_{j=1}^{M} \left(\frac{B(j)}{\mu_i}\right)^{N_j} \frac{B(i)}{\mu_i}$$

$$P[s] = K \prod_{j=1}^{M} \left(\frac{B(j)}{\mu_j}\right)^{N_j} \frac{B(j)}{\mu_j} \qquad (6\text{-}93)$$

The last term in each of the above two expressions arises from having a customer in service at the respective servers. Substituting these expressions into Equation (6-92) and simplifying, we get

$$\sum_{i=1}^{M} B(i) P_{ij} = B(j) \qquad (6\text{-}94)$$

which is what we postulated in Equation (6-90).

Now that we have Equation (6-94), we can generate a set of equations that we can solve simultaneously by setting one of the B terms equal to 1. In a manner similar to the previous example, we can also find the normalization constant K and therefore solve Equation (6-91) for the system's steady state probabilities, and also for the expected queue lengths using Equation (6-85).

If we consider the closed network over a long period of time, the relative throughput terms can be thought of as indicators of the relative number of times a customer visits the associated server, also called the "visit ratio." This interpretation is useful for determining which server is the most utilized, also known as "bottleneck analysis." Define the relative amount of work done by a server i as

$$\text{relative work by server } i = \frac{B(i)}{\mu_i} \quad (6\text{-}95)$$

Since this value is also the relative utilization of that server, the server with the highest such ratio is the system bottleneck.

Open Networks

Next, we will discuss another class of queuing networks, those that contain sources and sinks. We will assume that customers may arrive at any queue from an outside source according to a Poisson process that is specific for that queue. We can think of these arrival processes as all originating from a single arrival process with branches, each with an associated branching probability. Figure 6-17 shows such an arrival process and a hypothetical open network with M queues and associated servers.

We also assume exponential service rates for all servers in the system. In this case, the aggregate arrival rate is equivalent to the sum of all of the individual arrival rates discussed earlier. If each individual arrival rate is defined as

Arrival rate of queue $i = \alpha_i$

the aggregate rate is given as

$$\lambda = \sum_{i=1}^{M} \alpha_i \quad (6\text{-}96)$$

and the branching probabilities as

$$P_{0i} = \frac{\alpha_i}{\lambda} \quad (6\text{-}97)$$

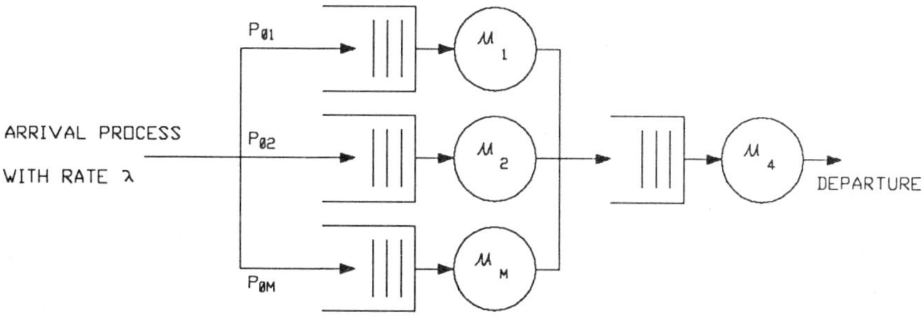

Figure 6-17 An arbitrary open system.

Customers leaving the system also do so with the probabilities defined as

$$P_{i0} = 1 - \sum_{j=1}^{M} P_{ij} \qquad (6\text{-}98)$$

This definition states that the probability of a job leaving the system is equal to the complement of the probability that a job will remain in the system.

As with the closed network discussed earlier, we can propose a set of throughput terms, denoted $B(i)$ for each queue and server i. Thus,

$$B(i) = \sum_{j=0}^{M} B(j) P_{ji}$$

$$B(i) = \sum_{j=1}^{M} B(j) P_{ji} + \alpha_i \qquad (6\text{-}99)$$

Since we know the thoughput arriving from the outside source, we can set

$$B(0) = \lambda \qquad (6\text{-}100)$$

and solve for the remaining B terms. In the case of an open network, the B terms will represent actual, not relative, throughput at a server i because they are derived from the aggregate arrival rate. Because of this, we can define each server's utilization as

$$u_i = \frac{B(i)}{\mu_i} \qquad (6\text{-}101)$$

After solving for all of the B terms, the steady state probabilities are given as

$$P(N_1, N_2, \ldots N_M) = K \prod_{i=1}^{M} \left(\frac{B(i)}{\mu_i}\right)^{N_i}$$

$$= K \prod_{i=1}^{M} u_i^{N_i} \qquad (6\text{-}102)$$

where again K is a normalization constant. We can sum all of the state probabilities and solve for K to obtain

$$K = \prod_{i=1}^{M} (1 - u_i) \tag{6-103}$$

Thus, the expression for the steady state probabilities becomes

$$P(N_1, N_2, \ldots N_M) = \prod_{i=1}^{M} (1 - u_i) u_i^{N_i} \tag{6-104}$$

If we look at Equation (6-45), we see that the expression just derived is actually the product of terms that can be obtained by treating each queue and server as an *M/M/*1 queue system in isolation. This result is known as "Jackson's theorem" and it states that, although the arrival rate at each server in an open system may not be Poisson, we can find the probability distribution function for the number of customers in any queue as if the arrival process were Poisson [and thereby use Equation (6-45) for the *M/M/*1 system]. Jackson's theorem further states that each queue system in the network behaves as an *M/M/*1 system with arrival rate defined by

$$\lambda_i = \alpha_i + \sum_{j=1}^{M} P_{ji} \lambda_j \tag{6-105}$$

which is simply (6-99) recast in more familiar terms.

It is worthwhile to note that Jackson's theorem applies to open systems in which the individual queue systems are *M/M/Ci*. That is, each server may actually be comprised of a different number (*i*) of identical servers. Thus, the steady state probabilities for each queue system in the network is also given by the equation for such a system in isolation with the arrival rate as described in Equation (6-105). The full proof of Jackson's theorem is given in [Jackson 1957].

ESTIMATING PARAMETERS AND DISTRIBUTIONS

Now that we have discussed various aspects of queuing theory, we should review some of the ways that we can parameterize the models that we choose. In this section, we will discuss various methods that can be used to determine whether a certain statistic or distribution appropriately describes an observed process. Specifically, we will cover hypothesis testing, estimators for some statistics, and goodness of fit tests. We will start with hypothesis testing in general.

A hypothesis test is a technique used to determine whether or not to believe a certain statement about a real world phenomenon and to give some measure as to what degree to believe the statement. A hypothesis is usually stated in two parts, the first concerning the statistic or characteristic that we are hypothesizing about and the second concerning the value that is postulated for the statistic. For example, we may hypothesize that the mean value of an observed process is less than 10, or that observed process is gaussian. The positive statement of a hypothesis as given above is usually called the "null" hypothesis and is denoted as $H0$. Associated with the null hypothesis is an alternative, denoted $H1$. The idea here is to have the two hypotheses complement each other so that only one will be selected as probable. The two hypotheses, $H0$ and $H1$, form the basis for the hypothesis test methodology that is outlined in the following paragraph.

A hypothesis test is usually performed in four general steps that lead to the acceptance or rejection of the initial hypothesis. The first step is to formulate the null hypothesis $H0$ and the alternative hypothesis $H1$. Next, decide upon a statistic to test against. The statistic is typically the sample mean or variance. Third, a set of outcomes for the test statistic is chosen so that the outcome of the test statistic will fall within the set with a specific probability, given that $H0$ is true. That is, if $H0$ is true, we say that the value of the test statistic will fall within the set selected (sometimes called the critical region) with probability P (also called the test's "level of significance"). The idea is to select a critical region so that the probability of the test statistic value falling within the region is small, typically between 0.01 and 0.05. An occurrence of this event, then, indicates that the hypothesis $H0$ is not a good choice and should be rejected. Conversely, we could select a large probability, say 0.9, in which case the occurrence of the event indicates that the null hypothesis should be accepted. The final step in the process is to collect some sample data and to calculate the test statistic.

The next immediate problem for performing a hypothesis test as described above is to define the expressions that describe the sample statistics we are interested in. These are commonly referred to as "estimators" because they estimate the statistic that could be derived from a distribution that exactly models the real process. The most commonly used estimators are the sample mean and the sample variance.

In order to calculate the sample statistics, we must first obtain a random sample from the experimental population. A random sample is defined here as a sequence of observations of the real world process, where each value observed has an equal probability of being selected and where each observation is independent of the others in the sample. Thus, a random sample is a sequence of random variables that are independent and identically distributed.

For a random sample of size n, where n is the number of samples obtained, the sample mean is defined as

$$\overline{X} = \sum_{i=1}^{n} \frac{X_i}{n} \qquad (6\text{-}106)$$

The sample variance is defined as

$$s^2 = \sum_{i=1}^{n} \frac{(X_i - X)^2}{n-1} \tag{6-107}$$

The sample standard deviation is defined as it was for the standard deviation of a distribution and is repeated here as

$$s = \sqrt{s^2} \tag{6-108}$$

In the above three expressions, the random variable X_i represents the ith observation in the random sample.

Now that we can calculate the statistics for a random sample of some phenomenon, how can we relate these estimates to the actual statistics of the underlying process? For this, we use a theorem known as the "sampling theorem." It states that, for a random sample as described above, with a finite mean, the sample mean and expected value are equivalent and that the sample variance and the variance are also equivalent. That is, the sample statistics are said to be consistent, unbiased estimators. It also states several other important relations, including the following expression relating the variance of the sample mean and the variance of the random variable describing the process. This expression

$$\text{VAR}[\overline{X}] = \frac{\text{VAR}[X]}{n} \tag{6-109}$$

states that as the sample size gets larger, the variance of the sample mean gets smaller, indicating that it is closer to the true mean of X.

The above estimates lead to still another question: Given that we know (or think that we know) the type of distribution that our random sample comes from, how do we estimate the parameters of such a distribution from the random sample data? There are two widely used methods for doing just this, the method of moments and maximum likelihood estimation.

The method of moments is useful when we think we know the distribution of the sample but do not know what the distribution parameters are. Suppose the distribution whose parameters we wish to estimate has n parameters. In this method, we first find the first n distribution moments as described earlier in Chapter 4. Next, we calculate the first n sample moments and equate the results to the moments found earlier. From this we get n equations in n unknowns, which can then be solved simultaneously for the desired parameters. We derive the kth sample moment for a sample size of m samples as

$$M_k = \sum_{i=1}^{m} \frac{X_i^k}{m} \tag{6-110}$$

where X_i is the i sample point in the random sample.

In maximum likelihood estimation, we try to pick the distribution parameters that maximize the probability of yielding the observed values in the random sample. To do this, we first form what is called the "likelihood function." This consists of the values of the assumed probability distribution function at the points observed in the random sample. This function, for a continuous random variable whose distribution has only one parameter, is

$$L = f(x_1)\, f(x_2)\, f(x_3)\, \ldots\, f(x_m) \tag{6-111}$$

For a random variable whose distribution has n parameters, we will have n equations like Equation (6-111). We then find the maximum of each equation with respect to each parameter. Finally, the set of n equations in n unknowns can be solved for the necessary parameters.

Now that we have outlined several methods for estimating the statistics of a distribution that describes the real world process, we turn our attention to the reliability of our estimates. One measure of this reliability is called the confidence interval. A confidence interval is defined as a range of values, centered at the estimate of the statistic of interest, where the actual value of the statistic will fall with a fixed probability. For example, a 90 percent confidence interval for the mean of a particular random variable based upon a given sample may be defined as the range of values within a distance r of the estimated mean. In this case, r is chosen so that the fraction of times that a sample lands within the interval is 90 percent. The general procedure for defining a confidence interval requires the construction of a known distribution, say C, from the estimates of the statistic being estimated. Next, we pick an interval so that

$$P(a < C < b) = z \tag{6-112}$$

where z is the desired confidence level. Finally, we evaluate C using the values X_i so that the relationship

$$a < C(X_i) < b \tag{6-113}$$

is maintained. We can alternatively solve C for the points X_a and X_b where $C(X_a) = a$ and $C(X_b) = b$. These are the end points of the $100z$ percent confidence interval.

The above procedure assumes that we know the distribution of C before we find the confidence interval. If this is not the case, and the sample size is large, we can assume that the sample distribution is normal and can obtain a reliable confidence interval for the value of the mean. In this case, we first form the statistic

$$T = \frac{(X - \mu)}{\sigma/\sqrt{n}} \tag{6-114}$$

Since X is assumed normal, T in this case is also normal with a mean of 0 and a standard deviation of 1. As above, we define a percent confidence interval and determine a and b so that

$$P(a < T < b) = z \qquad (6\text{-}115)$$

The desired confidence interval for the mean is then given by

$$X - \frac{b\sigma}{\sqrt{n}} < \mu < X + \frac{a\sigma}{\sqrt{n}} \qquad (6\text{-}116)$$

Confidence intervals for the variance when the population distribution is unknown can be found using the above method, although the results will be poor if the actual population distribution is far from normal.

Now that we have explored several techniques for estimating the parameters of distributions, we will look at some methods for finding a distribution that fits the sampled data. Typically, we will have found the sample mean and standard deviation and now want to find a random variable that adequately represents the sample population. The tests employed here are usually called "goodness of fit" tests. We will discuss two tests, the chi-square test and the Kolmogorov-Smirnov test. These tests fall under the general heading of hypothesis testing and, therefore, we use the same hypothesis-forming techniques described earlier. In both tests, we start with a null hypthesis that the population has a certain distribution, and then we obtain a statistic that indicates whether we should accept the null hypothesis.

In the chi-square test, we determine whether the distribution of the null hypothesis appropriately fits the population by comparing the catagories of the collected sample value to what can be generated by the assumed distribution. The premise is that we can find k bins, $B_1...B_k$, so that each value in the random sample falls into one and only one bin. After finding an appropriate set of bins, we partition the samples into them and record the number of samples that land in each. Next, we take a corresponding number of samples from the hypothesized population distribution and allocate them to the same bins found above. If any of the second set of samples, those taken from the distribution, fail to fall in only one bin, we have not selected an appropriate set of bins and must choose another set. For whatever type of distribution that we are testing against, the appropriate distribution parameters can be found using one of the estimation techniques described earlier. Continuing with the test, we now calculate the following statistic:

$$C = \sum_{i=1}^{k} \frac{(NS_i - ND_i)^2}{ND_i} \qquad (6\text{-}117)$$

where NS_i denotes the number of elements in bin i due to the random sample, and ND_i is the number in bin i due to the hypothesized distribution. The basis of this test is that the statistic of Equation (6-117) has a chi-square distribution. The degree of freedom of the chi-square distribution is defined as one less than the number of sample bins minus the number of parameters in the hypothesized distribution.

$$m = k - 1 - \text{number of parameters} \tag{6-118}$$

Next, we decide on the level of significance that we wish to test for. Using the following expression, we can calculate the probability density function for a chi-square distribution with n degrees of freedom.

$$f_X(x) = \begin{cases} \dfrac{1}{\left(\dfrac{n}{2} - 1\right)!} (2^{-n/2}) (x^{\frac{n}{2} - 1}) (e^{-\frac{x}{2}}) & \text{for } x > 0 \\ 0 & \text{otherwise} \end{cases} \tag{6-119}$$

The final step is to find the value of X for which the integral with respect to x of Equation (6-119), evaluated from x to infinity, is equal to the desired level of significance. The final test states that if the value of x just found is greater or equal to the chi-square statistic calculated in Equation (6-117), the assumed distribution is not a good fit at the desired level of significance. That is, we reject the null hypothesis if

$$x \geq C$$

An alternative approach for the chi-square test is to form the value $C - \varepsilon$, where ε is some small value. We then use the result to find the probability that x is greater than $C - \varepsilon$. The resultant probability gives us an indication as to the approximate level or significance that we may accept the null hypothesis. Several references ([Allen 1978], and [Trivedi 1982]) give tables for the critical values of the chi-square distribution. These tables may be used in place of calculating the distribution values as described above.

Another so-called goodness of fit test is the Kolmogorov-Smirnov test. The test is based upon the magnitude ordering of the sample, the calculation of the maximum difference between the sample points and the assumed distribution, and a determination of the level of fit of the assumed distribution. A formal description of this test appears in a number of statistics texts. Here we will describe a more intuitive approach that is somewhat easier to experiment with.

As mentioned earlier, the first step of this test is to arrange the sample values in ascending order according to magnitude. For each point x_i in the arranged sample, we find the fraction f_i of the number of total samples that are less in magnitude than the given value. Next, for the assumed distribution, we find the value K_i that will yield the same fraction f_i for a given number of samples. Finally, we plot K_i verses x_i for all i. The resulting plot will indicate a good fit if the data forms approximately a straight line with a slope of unity. If the fit is a straight line with a slope other than unity, the assumed distribution parameters may be tuned to achieve the desired results. Otherwise, we should try another assumed distribution.

SUMMARY

The areas covered in this chapter, from stochastic processes to queuing theory to basic estimation, span a wide range of topics, each with a wealth of specialities and techniques. The treatment given herein, although brief, is intended to illustrate the usefulness of statistical analysis and queuing theory and to provide a basis for understanding some of the techniques and methods used in simulation. More detailed discussions of the issues associated with basic probability and statistics are found in many basic probability texts, notably [Walpole 1978], [Papoulis 1984] and, for a queuing theory slant, in [Allen 1978]. Queuing theory topics are discussed in [Allen 1978], and also in [Trivedi 1982], [Lavenberg 83] and the classics [Kleinrock 1975, 1976]. Estimation, as related to queuing systems, is treated in [Trivedi 1982] and [Allen 1978].

REFERENCES

Allen, A.D. *Probability, Statistics and Queuing Theory with Computer Science Applications*, Academic Press Inc., Orlando, FL, 1978.

Kleinrock, L. *Queuing Systems: Volume I; Theory*, Wiley and Sons, New York, NY, 1975.

Kleinrock, L. *Queuing Systems: Volume II; Computer Applications*, Wiley and Sons, New York, NY, 1975.

Jackson, J. R. *Networks of Waiting Lines,"* Operations Research, August 1957.

Lavemberg, S. S., ed. *Computer Performance Modeling Handbook*, Academic Press Inc., New York, NY, 1983.

Papoulis, A. *Probability, Random Variables, and Stochastic Processes*, McGraw-Hill Book Company, New York, NY, 1984.

Trivedi, K. *Probability and Statistics with Reliability, Queuing and Computer Science Applications*, Prentice-Hall, Englewood, NJ, 1982

Walpole, R. E. *Probability and Statistics for Engineers and Scientists*, Macmillan Publishing Company, New York, NY, 1978.

7

Computational Methods for Queuing Network Solutions

In Chapters 4 and 6, we introduced probability theory and analysis techniques for performing classical queuing system analysis. Those analyses, however, tend to be complex even for simple systems. In an effort to rectify this situation, three alternative analysis methods have emerged.

The first, from Buzen, gives a technique for finding the normalization constant that is required for the solution of certain product form networks [Buzen 1973]. The method does not require the solution of the normalization summation as described in Chapter 4. Instead, it uses an iterative solution that is simpler to implement.

The next, from Denning and Buzen [Denning 1978], introduced a methodology for assessing the match of a given assumption for the system under analysis. In addition, they defined the performance quantities of interest in terms of their operational relationships in the system under study. For that reason, this kind of analysis is known as "operational analysis."

The third analysis method, from Reiser and Lavenberg [Reiser 1978], attempts to simplify the analysis of queuing networks. By using the mean waiting time and mean queue size, in conjunction with Little's result, the solution of a system of queues can asymptomatically approach the exact solution, with simpler computational requirements. This type of analysis is called "mean value analysis."

In this chapter, we will discuss the methods and models mentioned above. Some of the results are specific to the type of model that is used, others are more general. The specific model cases, however, can be used to approximate a given communication system or portion of a system and to obtain an initial feeling for the system's actual behavior.

CENTRAL SERVER MODEL

The central server model, shown in Figure 7-1, was originally proposed as a model for jobs in a multiprogramming computer. It is a closed network and we assume that a constant (*K*) number of jobs are always in process. In the original model, programs are serviced by the CPU (server 1) and then are routed to one of *M* - 1 I/O devices (servers 2 through *M*). After receiving I/O service, the program again queues for CPU time. If a program completes execution, it is rerouted into the CPU queue to start another job, thereby keeping the number of jobs in the system equal to *K*. This can be thought of as a system in which there is always a job waiting to enter the system at the CPU queue, but it will not do so until a job completes. The actual jobs in the system, therefore, may vary over time, but the number in circulation at any given time remains constant.

The central server model can be adapted to represent other systems besides a CPU and its associated I/O devices. For instance, we could choose server 1 to represent a multichannel DMA controller and servers 2 through *M* to represent the output channels. Or, we could adjust the branching probabilities to represent a system in which the jobs remain constant and never complete (i.e., $P_1 = 0$). This could be useful for a dedicated I/O server. Alternatively, we could choose one of the servers 2 through *M* to represent an idle period for a job or I/O channel.

Although we may be able to formulate a central server model that somewhat reflects the actual situation, the match may not be precise. The benefit of this model, however, is the computational simplicity of many of its important performance parameters. Next, we will develop the computational model for this queuing network.

In the central server model, the servers are assumed to have exponential service time distributions. As shown in Figure 7-1, the exit from the central server has several branches, each with an associated branching probability P_i. There are a total of *K* customers (jobs) in the system at any time. Let us define the state of the system as we did in Chapter 6:

$$S = (k_1, k_2, \ldots k_n) \tag{7-1}$$

where k_i denotes the number of customers in queue *i*. Thus

$$\sum_{i=1}^{M} k_i = K \tag{7-2}$$

If we define $B(i)$ as the probability of going to server *i* after service at server 1, and we let $B(1) = 1$, then

$$B_i = P_i \quad \text{for } i = 2, 3, \ldots M \tag{7-3}$$

Using the same techniques as shown in Chapter 6 for closed queuing networks, we can obtain the state probabilities as

Computational Methods for Queuing Network Solutions 165

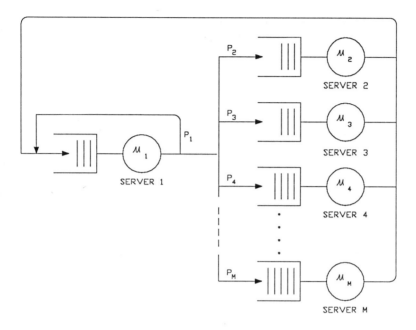

Figure 7-1 Central server model.

$$P(k_1, k_2, \ldots k_M) = \text{norm} \prod_{i=2}^{M} \left(\frac{P_i}{\mu_i}\right)^{k_i} \left(\frac{1}{\mu_1}\right)^{k_1} \tag{7-4}$$

This equation for the state probabilities is called "product form," and it can be solved by finding the normalization constant, Norm, as outlined in Chapter 6. The Chapter 6 methods, however, require the solution of M simultaneous equations. An alternative method with fewer computations is outlined below.

Let

$$G(k) = \sum_{\text{all states } j} P(j)/\text{norm} = \frac{1}{\text{norm}} \tag{7-5}$$

$$G(k) = \sum_{\text{all states}} \prod_{i=1}^{M} \left(\frac{P_i \mu_1}{\mu_i}\right)^{k_i} \tag{7-6}$$

with the following definition

$$\left(\frac{P_1 \mu_1}{\mu_1}\right)^{k_1} = 1 \tag{7-7}$$

Let the states for the summation include all states where Equation (7-2) holds. Also, we stipulate that $k_i \geq 0$. Define another function $g(k,m)$ where there are m queues in the system instead of M. The following, then, is true:

$$g(k, m) = G(k) \tag{7-8}$$

where k is the total number of customers in the system with M queues. Thus, we can further define $g(k,m)$ as

$$g(k, m) = \sum_{\text{all states}} \prod_{i=1}^{m} \left(\frac{P_i \mu_1}{\mu_i}\right)^{k_i} \tag{7-9}$$

We can break up the right-hand summation as

$$g(k, m) = \sum_{\substack{\text{all states} \\ \text{with } k_m > 0}} \prod_{i=1}^{m} \left(\frac{P_i \mu_1}{\mu_i}\right)^{k_i} + \sum_{\substack{\text{all states} \\ \text{with } k_m = 0}} \prod_{i=1}^{m} \left(\frac{P_i \mu_1}{\mu_i}\right)^{k_i} \tag{7-10}$$

For the first summation in Equation (7-10), if we always have at least one customer in queue m, we can think of the system as having $k - 1$ customers circulating through m queues. We must also remove the product term that relates to customers in queue m. Similarly, in the second summation, if queue m is always empty, then we can think of the system as having $m - 1$ queues and k customers. Thus, Equation (7-10) becomes

$$g(k, m) = \frac{P_m \mu_1}{\mu_m} \sum_{\substack{\text{all states} \\ \text{with } k-1 \text{ customers}}} \prod_{i=1}^{m} \left(\frac{P_i \mu_1}{\mu_i}\right)^{k_i} + \sum_{\substack{\text{all states} \\ \text{with } m-1 \text{ queues}}} \prod_{i=1}^{m} \left(\frac{P_i \mu_1}{\mu_i}\right)^{k_i} \tag{7-11}$$

The two summations can be rewritten, using Equations (7-6) and (7-8), as

$$g(k, m) = \frac{P_m \mu_1}{\mu_m} g(k - 1, m) + g(k, m - 1) \tag{7-12}$$

For $k = 0$, and for $m = 1$, Equation (7-12) becomes

$$g(0, m) = 1 \quad \text{for } m = 1, 2, \ldots M \tag{7-13}$$

$$g(k, 1) = 1 \quad \text{for } k = 0, 1, \ldots K \tag{7-14}$$

We now have a set of initial conditions [Equations (7-13) and (7-14)] and a recursive relationship Equation (7-12) for calculating the values up to $g(K,M) =$

Computational Methods for Queuing Network Solutions

$G(K)$. Then, we can use Equations (7-5) and (7-4) to calculate the state probabilities. The computation is illustrated below with an example.

Suppose we have a network like that shown in Figure 7-1 where $M = 3$, $\mu_1 = 0.9$, $\mu_2 = 0.5$, $\mu_3 = 0.9$, $P_1 = 0.7$, $P_2 = 0.2$, and $P_3 = 0.1$. Furthermore, suppose that there are $k = 2$ customers in the system. From Equation (7-13), we know that

$g(0,1) = 1$

$g(0,2) = 1$

$g(0,3) = 1$

and from Equation (7-14) we know that

$g(0,1) = 1$

$g(1,1) = 1$

$g(2,1) = 1$

We can arrange these values in a grid as shown in Figure 7-2. The computation proceeds one row at a time and ends up with a value for $g(2,3) = G(2)$.

For example, to calculate $g(1,2)$, we would proceed as follows

$$g(1,2) = \frac{P_2 \mu_1}{\mu_2} g(0,2) + g(1,1)$$

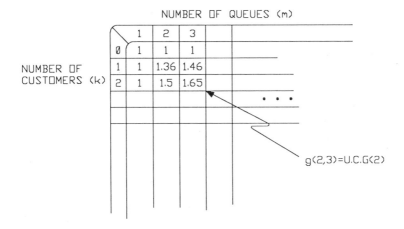

Figure 7-2 Normalization constant matrix.

$$g(1,2) = \frac{(0.2)\,(0.9)}{0.5}(1) + 1$$

$$g(1,2) = 1.36$$

Thus, for Figure 7-2, the normalization constant is equal to 0.6. With two customers, we can now calculate the state probabilities using Equation (7-4).

Buzen gives several expressions for performance measures that are based upon this general computational structure. One of these measures is the device utilization, U_i, for server i. Normally, we define the utilization of a device as the sum of the state probabilities where there is at least one customer at server i.

$$U_i = \sum_{\substack{\text{all states } i \\ \text{where } k_i > 0}} P(i) \tag{7-15}$$

$$U_i = \frac{1}{G(k)} \sum_{\substack{\text{all states } i \\ \text{where } k_i > 0}} \prod_{i=1}^{m} \frac{(P_i \mu_1)^{k_i}}{\mu_i} \tag{7-16}$$

Using similar reasoning as we did for Equation (7-11), we can treat Equation (7-16) as a system with one less customer, multiplied by the factor that accounts for always having a customer at queue i. Thus we get

$$U_i = \frac{1}{G(k)} \frac{P_i \mu_1}{\mu_i} \sum_{\substack{\text{all states } i \\ \text{with } k-1 \text{ customers}}} \prod_{j=1}^{m} \frac{(P_j \mu_1)^{k_j}}{\mu_j} \tag{7-17}$$

so

$$U_i = \frac{P_i \mu_1}{\mu_i} \frac{G(K-1)}{G(K)} \tag{7-18}$$

We have already calculated the values for $G(K)$ and $G(K-1)$ as in Figure 7.2, so calculating the utilization of a device is straightforward. From this, we can find the throughput of device i as

$$\lambda_i = U_i \mu_i = P_i \mu_1 \frac{G(K-1)}{G(K)} \tag{7-19}$$

Looking back on (7-18), we can extend this expression to find the probability that the queue length at server i will be greater or equal to than some value n.

Computational Methods for Queuing Network Solutions 169

$$P(N_i \geq n) = \frac{1}{G(k)} \sum_{\substack{\text{all states} \\ \text{where } k_i \geq n \text{ customers}}} \prod_{j=1}^{m} \left(\frac{P_j \mu_1}{\mu_j}\right)^{k_j} \quad (7\text{-}20)$$

so that

$$P(N_i \geq n) = \left(\frac{P_i \mu_1}{\mu_i}\right)^n \frac{G(K-n)}{G(K)} \quad (7\text{-}21)$$

Applying Equation (7-21) to the case where the queue length is $n + 1$, we can obtain the probability of n customers in queue i.

$$P(N_i \geq n) = \left(\frac{P_i \mu_1}{\mu_i}\right)^n \frac{G(K-n)}{G(K)} - \left(\frac{P_i \mu_1}{\mu_i}\right)^{n+1} \frac{G(K-n-1)}{G(K)} \quad (7\text{-}22)$$

$$P(N_i \geq n) = \left(\frac{P_i \mu_1}{\mu_i}\right)^n \frac{1}{G(K)} \left(G(K-n) - \frac{P_i \mu_1}{\mu_i} G(K-n-1)\right) \quad (7\text{-}23)$$

Now that we have derived an expression for the probability of having n customers in the queue, we can use Equation (4-54) to obtain an expression for expected queue length.

$$E[N_i] = \sum_{j=1}^{K} j P(N_i = j) \quad (7\text{-}24)$$

$$= \sum_{j=1}^{K} j \left[\left(\frac{P_i \mu_1}{\mu_i}\right)^j \frac{1}{G(K)} \left(G(K-j) - \frac{P_i \mu_1}{\mu_i} G(K-j-1)\right)\right]$$

$$= \frac{1}{G(K)} \left(\frac{P_i \mu_1}{\mu_i} G(K-1) - \frac{P_i \mu_1}{\mu_i} G(K-2)\right)$$

$$+ 2 \left(\left(\frac{P_i \mu_1}{\mu_i}\right)^2 (G(K-2) - \frac{P_i \mu_1}{\mu_i} G(K-3))\right)$$

$$+ \ldots + K \left(\frac{P_i \mu_1}{\mu_i}\right)^K (G(0) - \frac{P_i \mu_1}{\mu_i} G(-1)))$$

We can now expand and collect terms, keeping in mind that we have defined $G(K < 0) = 0$, to get

$$E[N_i] = \frac{1}{G(K)} \sum_{j=1}^{K} \frac{P_i \mu_1}{\mu_i} G(K-j) \qquad (7\text{-}25)$$

The expected delay through a queue, then, can be found from Little's result, using the throughput and expected queue length values just found.

$$E[w_i] = \frac{E[N_i]}{\lambda_i} \qquad (7\text{-}26)$$

The techniques given above allow the efficient computation of the important statistics for the model of Figure 7-1. Next, we will discuss another, slightly more general computational method, mean value analysis.

MEAN VALUE ANALYSIS

The analyses described so far have all calculated, at one point or another, expressions for queue length distributions or state probabilities. This is perfectly acceptable for rather simple systems involving few customers and queuing systems. In the following discussion, we will discuss an iterative method for finding some of the performance measures of interest without calculating the aforementioned distributions. The drawback to this is that the analysis refers only to the mean values of certain performance measures.

The techniques that we are interested in here apply to closed queuing networks that have a product form solution for the state probabilities [for example, see Equation (6-104)]. The solutions are based on the assumption that a customer, arriving at any queue in a closed system that is at steady state, experiences a wait in the queue that is equivalent to the steady state wait time for that queue with the arriving customer removed. Thus, the behavior of a network with one more customer is based upon the behavior before its arrival. This assumption leads to an iterative algorithm where the steady state performance characteristics for the system with $n + 1$ customers are derived from the characteristics with n customers, which are derived from a system with $n - 1$ customers, and so on down to one customer.

The general algorithm, then, allows us to compute average values for queue length, throughput, server utilization and wait time by starting with an expression for one customer in the system and working up to any number of customers.

The main theorem behind mean value analysis states that the mean wait time for customers at any server in the network is related to the solution of the same network with one fewer customers. In conjunction with the wait time, we apply Little's result to obtain the total network throughput and then apply it again to each individual server to get the average queue length at each server.

Computational Methods for Queuing Network Solutions 171

The expression for wait time related to a network with one fewer customers is given as

$$w(k) = \frac{1}{\mu}[1 + N_q(k-1)] \tag{7-27}$$

where μ is the mean service time required by a customer at the server. The quantities $w(k)$ and $N_q(k-1)$ denote the mean wait time for a system with k customers at the queue and the mean number of customers in the queue with $k-1$ customers in the system, respectively. The above expression holds for systems with first-come-first-serve queuing disciplines with single, constant rate, servers at each queue.

Next, we can apply Little's result to find the mean throughput for the network

$$\lambda_k = \frac{k}{\text{avg wait time}} = \frac{k}{\sum_{\text{all } i} \emptyset_i \, w(k)} \tag{7-28}$$

where $\emptyset$ is the visit ratio for the server considering all other servers. The visit ratio values are explained a little later.

Finally, we can use Little's result on the servers to compare the average queue lengths.

$N_q(k)$ = arrival rate x average wait time

$$N_q(k) = \emptyset_i \, \lambda(k) \, w(k) \tag{7-29}$$

Now we have a new expression for mean queue length that we can use in Equation (7-27) to start another iteration.

The general procedure, then, is to start with an empty system ($K = 0$) and iterate Equations (7-27) through (7-29) until we reach the desired value of K. For one iteration, we calculate the values for each queue system in the network before passing on to the next equation. Figure 7-3 shows a simple network to illustrate the technique. In the example, if we start with 0 customers, we obtain the following quantities from Equations (7-27) through (7-29). In the following expressions, the subscripts denote the queue/server pair that the measure is associated with. The general iteration algorithm is given below:

The first iteration is

$$w_1(1) = \frac{1}{\mu_1}(1 + 0) = \frac{1}{\mu_1}$$

$$w_2(1) = \frac{1}{\mu_2}(1 + 0) = \frac{1}{\mu_2}$$

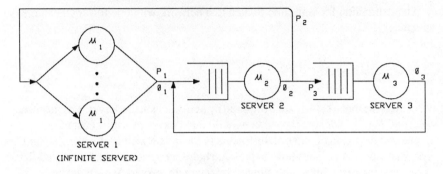

Figure 7-3 Network for mean value analysis.

$$w_3(1) = \frac{1}{\mu_3}(1+0) = \frac{1}{\mu_3}$$

$$\lambda(1) = \frac{1}{\varnothing_1 \mu_1 + \varnothing_2 \mu_2 + \varnothing_3 \mu_3}$$

$$N_1(1) = \varnothing_1 \lambda(1) w_1(1)$$

$$N_2(1) = \varnothing_2 \lambda(1) w_2(1)$$

$$N_3(1) = \varnothing_3 \lambda(1) w_3(1)$$

The second iteration is

$$w_1(2) = \frac{1}{\mu_1}(1+N_1(1))$$

$$w_2(2) = \frac{1}{\mu_2}(1+N_2(1))$$

$$w_3(2) = \frac{1}{\mu_3}(1+N_3(1))$$

$$\lambda(2) = \frac{2}{\varnothing_1 w_1(2) + \varnothing_2 w_2(2) + \varnothing_3 w_3(2)}$$

$$N_1(2) = \varnothing_1 \lambda(2) w_1(2)$$

$$N_2(2) = \varnothing_2 \lambda(2) w_2(2)$$

$$N_3(2) = \varnothing_3 \lambda(2) w_3(2)$$

The visit ratios, $\emptyset_i$, are obtained as follows. Pick a server and set its visit ratio value $\emptyset_i$ to 1. Next, formulate the equations that contribute to the visit ratio for that queue by looking at all queues that feed it. Equate the feeder visit ratios, multiplied by the respective branching probabilities, to the next in line ($\emptyset_i$). Continue this process for each queue system until we have m relationships in m unknowns, where m is the number of queuing systems. We can then solve this system of equations to obtain the desired visit ratios. Note that the visit ratios are relative quantities. For Figure 7-3, the visit ratios would be calculated as follows

$$\emptyset_1 = 1$$

$$\emptyset_1 = P_2 \emptyset_2 ; \emptyset_2 = \emptyset_1/P_1$$

$$\emptyset_2 = \emptyset_3 + \emptyset_1 ; \emptyset_3 = \emptyset_1 - \emptyset_{2_1}$$

The algorithm is iterated until we reach the desired network population, where we can calculate the mean performance measures for the network. Additional expressions for Equations (7-27) through (7-29) are given in [Reiser 1980] for other queuing disciplines and server types. Also, networks with multiple customer classes and multiple-server queuing systems are also addressed there.

OPERATIONAL ANALYSIS

The methods for performing queuing analysis given in Chapter 6 provide close approximations to the actual systems that they represent. It is contended however, that the assumption that the various distributions and relationships that we use to represent their real world counterparts cannot be proven to be absolutely accurate. Furthermore, the stochastic models studied earlier yield relationships that cannot be absolutely proven to be valid during any observation period.

Operational analysis, on the other hand, is based upon the observation of basic, measurable quantities that can then be combined into operational relationships. Furthermore, the observation period for which the system is analyzed is finite. The assumption is that the basic quantities, called operational variables, are measurable (at least in theory). The basic operational variables that are commonly found in operational analysis are listed below.

T = the observation period length   (7-30)

A = the number of arrivals during the observation period   (7-31)

B = the server busy time during the observation period   (7-32)

C = the number of job completions during the observation period

(7-33)

174   Modeling and Analysis of Local Area Networks

In addition to the measurable quantities, there are several fundamental relationships that define other useful quantities. These are listed below.

λ = arrival rate = A/T (7-34)

X = completion rate = C/T (7-35)

U = server utilization = B/T (7-36)

S = mean service time per job = B/C (7-37)

Several operational identities are also defined that hold under certain operational assumptions. The first, which relates server utilization to the completion rate and mean service time, is derived as follows

$$Xs = \left(\frac{C}{T}\right)\left(\frac{B}{C}\right)$$

$$= \frac{B}{T}$$

$$Xs = U \tag{7-38}$$

This relationship holds in all cases and is thus referred to as an operational law.

If we assume that the system is in steady state equilibrium, we can state that the number of arrivals and the number of completions during a given observation period will be equal (i.e., the flow balance assumption). This statement may not always be true, but it can be measured and verified for any observation period of interest. Thus, it is called an operational theorem. From this, we can derive another relationship that holds when the system is in flow balance.

$A = C$

$A/T = C/T$

$\lambda = X$

$\lambda s = Xs$

$\lambda s = U$ (7-39)

One advantageous property of operational analysis is that the technique can be used for open and closed networks. The one condition, however, that must be met is that the network under consideration must be what is called "operationally connected." That is, no server may be idle during the entire observation period.

Computational Methods for Queuing Network Solutions 175

For a closed system, we know the number of jobs in circulation in the network and we find the system throughput at a particular point in the network. Other quantities can then be derived using that throughput as a starting point. In an open system, the throughput at the network exit is assumed to be known and we use this to find the number of customers at the queues.

Let's look now at some basic operational quantities. Suppose that we have an arbitrary network that is observed over a period of T. For each queue system in the network, we observe and collect the following data

A_i = number of arrivals at queuing system i

B_i = busy time of server i

C_{ii} = number of times a job goes directly from server i to server j's queue

Jobs that arrive from an external source or that leave to an external sink are denoted by A_{0i} and C_{i0}. The number of arrivals to and departure from the system are given by

Number of arrivals $\quad A_0 = \sum_{j=1}^{m} A_{0j}$ (7-40)

Number of departures $\quad C_0 = \sum_{i=1}^{m} C_{i0}$ (7-41)

and the total number of completions at server i is given as

$$C_i = \sum_{j=1}^{m} C_{ij}$$ (7-42)

From the basic measured quantities defined above, several other performance quantities can be derived.

Utilization of server i: $\quad U_i = B_i/T$

Mean service time of server i: $\quad S_i = B_i/C_i$

Output rate of server i: $\quad Xi = C_i/T$

Routing frequency from server i to j: $\quad q_{ij} = C_{ij}/C_i$

176 Modeling and Analysis of Local Area Networks

We can represent the job completion rate of such a system as

$$X_0 = \sum_{i=1}^{m} X_i \, q_{i0} \tag{7-43}$$

and the utilization of any server i as

$$U_i = X_i \, s_i \tag{7-44}$$

If we think of the wait time at a particular server i at each increment of time during the observation period as the sum of the service times of the customers ahead of the new arrival, the total wait time accumulated for all jobs in the system over the period is

$$w_i = \int_0^T n_i(t) \, dt \tag{7-45}$$

The average queue length at the server in question is given as

$$N_i = \frac{w_i}{T} \tag{7-46}$$

and the response time of the server system as

$$R_i = \frac{w_i}{C_i} \tag{7-47}$$

Combining Equations (7-46) and (7-47), we obtain the operational equivalent of Little's result.

$$N_i = \frac{w_i}{C_i} \frac{C_i}{T} = R_i \, X_i \tag{7-48}$$

If the system under study is in steady state so that we have flow balance, we assume that the arrival rate to a queuing system is equal to the completion rate out of that same system. We can also derive the server throughput rate for any server j as

$$C_j = \sum_{i=0}^{m} C_{ij}$$

$$C_j = \sum_{i=0}^{m} \frac{C_i C_{ij}}{C_i}$$

$$C_j = \sum_{i=0}^{m} C_i q_{ij}$$

From Equation (7-35), we obtain the same expression as stated in Equation (7-43), but generalized for any server it is

$$X_j = \sum_{i=0}^{m} X_i q_{ij} \quad \text{for } j = 0, 1, \ldots m \tag{7-49}$$

The relation derived above yields a unique solution if applied to an open system because the input throughput, X, is known. In a closed system, Equation (7-49) will yield relative throughput rates because we do not know the absolute value of X_0.

Buzen defines the visit ratio as for server i (V_i) as the number of times a particular server i will be visited, relative to a given number of inputs. We can express this quantity as the ratio of the throughput at server i to the total input throughput.

$$V_i = \frac{X_i}{X_0} \tag{7-50}$$

If we assume that the flow of jobs in the network is balanced, we can set $V_0 = 1$ (since all jobs pass through the network input) and solve for all of the other visit ratios using the following expression:

$$V_j = q_{0j} + \sum_{i=1}^{m} V_i q_{ij} \tag{7-51}$$

Also, knowing the throughput of any server in the network allows us to find the throughput of any other server through a combination of Equations (7-51) and (7-50).

Now let's look at the total time a job remains in the system as a function of each server's throughput and average queue length. The total wait time for any job arriving at any server depends on how many jobs are ahead of the new one in the queue and upon the rate that jobs get completed by the server. At each server, then, we can use Little's result [Equation (7-48)] in combination with Equation (7-50) to obtain

$$\frac{N_i}{X_0} = V_i R_i \tag{7-52}$$

If we then sum Equation (7-52) over all servers in the network, we obtain a general expression that can be interpreted as Little's result applied to the total system.

$$\sum_{i=1}^{m} \frac{N_i}{X_0} = \sum_{i=1}^{m} V_i R_i \tag{7-53}$$

where the number of jobs in the system at any time is simply the sum of all jobs at the network's servers:

$$N = \sum_{i=1}^{m} n_i \tag{7-54}$$

So we have

$$\frac{N}{X_0} = \sum_{i=1}^{m} V_i R_i \tag{7-55}$$

The left-hand side of Equation (7-55) can be thought of as an application of Little's result to the system as a whole; thus, we define the system response time as

$$R = \frac{N}{X_0} = \sum_{i=1}^{m} V_i R_i \tag{7-56}$$

The final topic that we will cover under operational analysis is bottleneck analysis in a closed system. In every network, one of the queuing systems will eventually be unable to keep up with increased service demands as the number of jobs in the network increases. This particular server will subsequently determine the maximum throughput rate of the network as a whole. A device is considered to be saturated (e.g., unable to process jobs any faster) if its utilization becomes one. In this case, the throughput will be inversely proportional to the service time since there will always be a job in service.

$$X_i = \frac{1}{s_i} \tag{7-57}$$

If we combine (7-44) and (7-50), we can express the relative utilization of any two servers in the network as follows

$$\frac{U_i}{U_j} = \frac{V_i S_i}{V_j S_j} \qquad (7\text{-}58)$$

Note that the ratios of the server utilizations do not depend upon the throughput of either server; the ratio remains constant independent of system load. Thus, the device with the largest value of $V_i S_i$ will become the network's bottleneck as load increases.

It is possible, then, to find the maximum possible system throughput when the bottleneck is in saturation. Since, for the bottleneck server b, throughput is equal to the inverse of the service time, we can combine Equations (7-57) and (7-50) to obtain the maximum system throughput:

$$V_b = \frac{X_b}{X_0} = \frac{1}{X_0 S_b}$$

$$X_0 = \frac{1}{V_b S_b} \qquad (7\text{-}59)$$

The network response time, in the saturation case, is given by Equation (7-56) as

$$R = \frac{N}{X_0} = N V_b S_b$$

and is thus limited by the bottleneck server.

Buzen and Denning extend the operational results discussed above to systems with load-dependent behavior in [Buzen 1978]. Also, an earlier proposal for operational analysis of queuing networks can be found in [Buzen 1976].

SUMMARY

The application of the techniques discussed in this chapter enable one to calculate, under certain assumptions and conditions, many of the interesting performance quantities that can be found with traditional queuing theory analysis. The results obtained, however, are often more intuitive and can be more easily related to the actual system for which they are intended.

REFERENCES

Buzen, J. P. "Computational Algorithms for Closed Queuing Networks with Exponential Servers," Communications of the ACM, September 1973.

Buzen, J. P. "Fundamental Operational Laws of Computer System Performance," Acta Informatica, Vol. 7, No. 2, 1976.

Buzen, J. P. and P. J. Denning, "The Operational Analysis of Queuing Network Models," ACM Computing Surveys, September 1978.

Reiser, M. and S. S. Lavenberg, "Mean-Value Analysis of Closed Multichain Queuing Networks," Journal of the ACM, April 1980.

8

Hardware Test Beds

In the previous chapters, we covered modeling from several perspectives ranging from simulation to queuing models to operational analysis. For those perspectives, only limited amounts of data are actually measured on an actual system. Often, the simplifying assumptions that are made so that model results are calculable enable us to obtain only an approximate analysis of the system's behavior. Also, the load conditions that are presented to an analytical or simulation model often are not tested in a real world situation. These factors have two ramifications. The first is that more detailed analysis is difficult because of the lack of adequate real world data. The second is that, even with a detailed model, validation of the model and its results must be weak at best. The latter statement is especially true for general-purpose simulation models such as those discusssed throughout this book. Before a simulation can be used to predict the performance of any system, the results of its execution must be compared against a known baseline, and the simulation must be adjusted accordingly. One method of achieving this is through the instrumentation and collection of performance data on an actual system. The results of these measurements are compared with the predicted results from a simulation model of the same system. When the results agree to within some predetermined tolerance, the model is considered validated.

This chapter discusses the use of prototype hardware test beds as a tool for ascertaining actual measures for some of the performance quantities of interest, for performing controlled experiments to determine the operational characteristics of different parts of a network, and for the validation of software simulation models. In particular, we will describe the implementation of a hardware test bed, define the measurable quantities that we are interested in, derive operational relationships for nonmeasured quantities, and give some results.

The construction of a special-purpose test bed can be costly if done solely for the purpose of estimating the final system performance. Often, however, a proof of concept prototype is constructed to test and validate design assumptions, to gain experience with the system, and to provide a vehicle for advanced development. Given that a prototype system often exists, it is advantageous to also consider instrumentation and test provisions in the prototype design. When performed within the scope of the prototyping effort, the relative cost of special performance measurement facilities becomes more acceptable. Some important facilities, which we will describe for a specific example later in this chapter, could include a systemwide time base for obtaining synchronous measurements, time-tagging hardware or software for time-stamping events, counters for recording the number of occurrences of important events, and scenario drivers that can inject a known load into the system being modeled. Of course, it is desirable to make these facilities as unintrusive as possible so that their usage does not interfere with the normal operation of the network under question. In some cases, portions of the final system software configuration may be substituted for by special purpose measurement facilities. The remainder of this chapter will discuss a prototype network configuration and will illustrate the techniques employed to measure its performance characteristics.

The network that we will be discussing is situated in a prototype test bed that is instrumented for data collection and can generate network traffic. Each test bed node contains a host controller that can emulate a known traffic load or generate any specified pattern of message traffic. Experiments can be repeated so that different measurements can be taken or so that a specific communication-related parameter may be varied. Thus, the prototype system's loading and recording mechanisms can be controlled in order to observe different network performance phenomena.

In constructing a prototype test bed such as the one discussed here, it is desirable to keep hardware development costs at a minimum, to provide a flexible system so that changes in network design can be accommodated, and to provide the general-purpose driver capabilities discussed above. One method of keeping hardware development costs down is to use off-the-shelf hardware components as much as possible. All node components that are not network specific can be implemented using standard board or system-level products. For example, the host processor for a node could be implemented with a single board computer or even with a personal computer.

Flexibility in the design of the network specific components is essential for minimizing the impact of the inevitable design changes that occur during the early network design and prototyping phase. One useful method for achieving a flexible prototype design is to reduce the speed of operation of the network. This allows some functions of the network to be implemented with more general-purpose components such as a programmable microcontroller of state machine. After the prototype design has been analyzed and a near-final configuration decided upon, these functions can be transitioned into higher-speed implementations. The assumption here is that a uniform scaling of the speed of operation across all network-sensitive components will yield results that can be scaled back up to reflect the actual system's performance. This may not hold true in the strictest sense, such as where hardware characteristics change at higher speeds, but it will generally hold if the functionality of the network as a whole does not change.

Hardware Test Beds 183

In order to provide general-purpose network driver and data collection capabilities, it is almost always necessary to have a detached host whose only function is to generate network traffic and collect results. Also, it may be necessary to design in additional resources whose only functions are to assist in traffic generation of data collection. It is important to adhere as much as possible to a layered network standard such as the International Standards Organization's model for Open System Interconnect (ISO's OSI model). By doing this, changes can be more or less localized to the level that is most affected, whereas the other levels can maintain their functionality. Thus, the same standards that provide a degree of interoperability among networks of different types also provide us with a useful template for building a flexible prototype system.

The hardware test bed used here, for example, consists of several network nodes connected with a token bus LAN. Each node contains two single-board computers, one that implements the simulated host functions (the host) and provides for network loading and data collection and one that provides high-level control functions for the network hardware (the input/output processor, or IOP). Additionally, each node contains a network adapter whose function is to implement the network protocol.

In this particular case, the network test bed models a general-purpose serial communication network. With a stable host and IOP design, a number of different network types can be implemented by using different network adapters and front-end hardware. The network that we will examine uses a token access protocol. In this protocol, the node that holds the token has the option to transmit data, if there is message traffic queued for transmission by the host processor. If the node does indeed have a message to send, it broadcasts the message over the communication bus. All other nodes listen for their identifying address in the message header and accept the message if it is destined for them. After the transmission of a message has been completed, or if there is no message to send, the token is passed to the next node on the network in a round robin fashion. The next node may or may not be physically adjacent to the current node.

The overall structure of the network test bed is shown in Figure 8-1. A number of nodes, each with a network adapter, are attached to a linear token bus and also to a data analysis computer. During a test run, the network bus is used to transfer the simulated load. At the completion of the test, each node transmits its collected data to the data analysis computer for synthesis and analysis. Each node has an architecture as shown in Figure 8-2.

The host computer serves two functions in this architecture. The first is to implement part of the layered protocol and to provide a simulated message load to it. The second is to collect the necessary performance data for subsequent analysis. Figure 8-3 shows the general structure of the host software that implements these functions.

The IOP controls the flow of message traffic onto and off of the network through the network adapter. It also controls the DMA channels, provides a standard

184 Modeling and Analysis of Local Area Networks

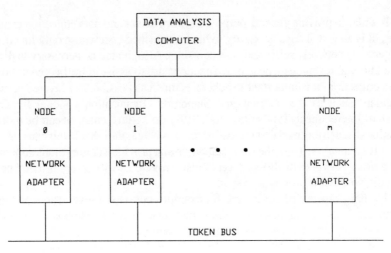

Figure 8-1 General test bed configuration.

interface to the host computer, and collects network-specific performance statistics. Figure 8-4 shows the IOP's functional architecture.

As mentioned earlier, it is advantageous to have the test bed components conform to a layered protocol standard. The test bed under discussion here implements levels 1 through 4, and part of level 5, of the OSI model for layered protocols. Figure 8-5 shows how the various components map the the standard. In

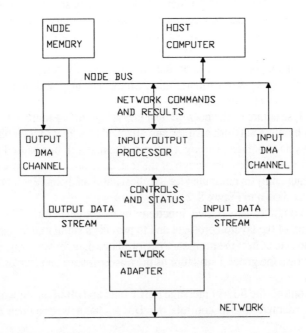

Figure 8-2 Test bed node architecture.

Hardware Test Beds 185

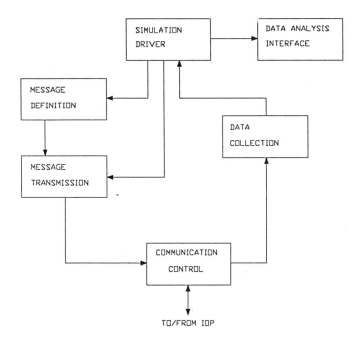

Figure 8-3 Host software architecture.

the layered model of Figure 8-5, the physical level implements the electrical and physical functions that are required to link the nodes. The data link layer provides the mechanisms necessary to reliably transmit data over the physical link. Level 3, the network level, controls the switching of data through the network. In networks with multiple transmission paths, the level 3 function controls which links a message will be transferred over. At the transport level, an error-free communica-

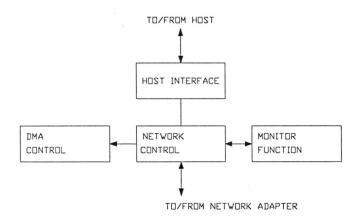

Figure 8-4 IOP functional architecture.

ISO OSI LEVEL	TEST BED
LEVEL 7: APPLICATION	SIMULATION DRIVER
LEVEL 6: TRANSFORMATION/PRESENTATION	NOT IMPLEMENTED
LEVEL 5: SESSION	PARTIALLY IMPLEMENTED IN HOST
LEVEL 4: TRANSPORT	IOP
LEVEL 3: NETWORK	
LEVEL 2: DATA LINK	NETWORK ADAPTER
LEVEL 1: PHYSICAL	

Figure 8-5 Test bed/ISO OSI correspondence.

tion facility between nodes is provided. Session control involves the initiation, maintenance, and termination of a communication session between processes. Level 6 provides any data translation, compaction, or encoding/decoding services that may be required by the application. At the top resides the application, which is any process that uses the commmunication facilities.

For the example network, levels 1 and 2 provide physical connection via coaxial cables, the serialization and packing of data, and the synchronization and detection of data onto and off of the network. Since the network discussed here is a global bus, there is no need for the switching functions of level 3. In cases where this is a factor, however, the function would be implemented in the IOP. Transport control is best implemented in the IOP because it relieves the host of performing error detection and retransmission and of managing individual message packets. Part of level 5 is implemented in the host so that messages can be assembled and queued for transmission to other nodes. Mechanisms for establishing interprocess connections are not implemented.

The network that we will study as an example requires the acknowledgment of each message packet from the receiver. A missing or bad acknowledgment results in the retransmission of the packet in error. Messages are addressed to logical process identifiers, which are mapped to each node upon initialization.

In the test bed model, a sequence of messages is treated as a series of time-ordered events. The event times are generated in the host according to a probability distribution that is representative of the desired loading characteristics. The time of message generation is recorded and collected for post-run analysis. As a message is transferred through the protocol layers and across the network, it is time-tagged, again for later analysis. In the following section, we will illustrate the use of these time tags and other collected data from a run, derive the performance evaluation parameters of interest, and show some experimental results that exemplify the techniques.

DERIVATION OF PERFORMANCE EVALUATION PARAMETERS

As mentioned earlier, the message traffic for the network under examination is generated in the host. A queue of messages awaiting transmission is implemented in the node memory shown in Figure 8-2. A message is, therefore, said to enter the network from the host processor and to exit through the same.

After entering the network, the message is broken into a series of packets, each of which is transmitted serially by the network adapter under the control of the IOP. Only one network adapter may have control of the bus at any one time (i.e., only one may transmit at a time). This serial access is controlled by the circulating token. Thus, the network represents a single-server queuing system where the service provided is the message transmission. All messages in the system are of the same priority so that the system has only one customer class.

Because access to the network is serial by node and because the only server in the system is the network itself, we can consider all message packet arrivals to the server as originating from a single queue. Thus, this single conceptual queue contains the combination of all messages in the individual message queues, ordered by time. Figure 8-6 illustrates this concept.

For this example, we will assume that messages arrive at the server according to a Poisson distribution. Thus, the probability that we get n arrivals at host i in an interval of length t is given as

$$P(n \text{ arrivals in interval t}) = \frac{(\lambda_i t)^n e^{-\lambda_i t}}{n!} \qquad (8\text{-}1)$$

where λ_i is the average interarrival rate at host i. For the Poisson distribution, the time between arrivals is exponentially distributed, and the interarrival time for messages at host i is generated as

$$A_i = -\frac{1}{\lambda_i} \ln x \qquad (8\text{-}2)$$

where X is a uniformly distributed random variable ranging from 0 to 1. The average interarrival time for the conceptual single network server, then, can be represented as

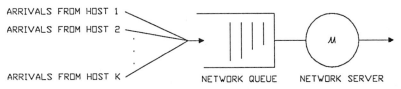

Figure 8-6 Conceptual network server.

$$A_i = -\sum_{i=1}^{K} \frac{1}{\lambda_i} \ln x \tag{8-3}$$

We can represent the state of the system during the observation period as the number of messages awaiting transmission through the network. Because of the property of the Poisson arrival process that the probability of no more than 1 arrival or completion in any time period approaches one as the interval length approaches 0, the state transitions satisfy the one-step assumption. That is, the system state only transitions to neighboring states. A state is denoted $n(t)$ and defines the number of message packets awaiting transmission at time t.

We will perform an analysis that is based upon the operational analysis techniques discussed in Chapter 7. The quantities for this evaluation are summarized below.

W   Waiting time for a message packet measured from arrival into the network queue until the completion of transmission

B   Busy time for the network defined as the total time that there is at least one message packet in the system

The quantities defined above are derived from measurements of three basic quantities measured by instrumentation hardware and software in the test bed. The basic measured quantities are:

$A(n)$ Number of arrivals into the sustem when there are n message packets in the system

$C(n)$ Number of completions when there are n message packets in the system

$T(n)$ Total amount of time when there are n message packets in the system

Define the total over all n of each of the above quantities as follows

$$A = \sum_{i=0}^{K-1} A(i) \quad \text{(arrivals)} \tag{8-4}$$

$$C = \sum_{i=1}^{K} C(i) \quad \text{(completions)} \tag{8-5}$$

$$T = \sum_{i=0}^{K} T(i) \quad \text{(observation period)} \tag{8-6}$$

Hardware Test Beds   189

In the above summations, *K* represents the largest number of message packets awaiting transmission during the observation interval. If we assume flow balance, the total number of arrivals will equal the total number of completions during the observation period. The waiting and busy time defined earlier can be defined in terms of these quantities as

$$W = \sum_{i=1}^{K} iT(i) \tag{8-7}$$

$$B = \sum_{i=1}^{K} T(i) = T - T(0) \tag{8-8}$$

Along with the above measures, we obtain three additional measures, message transmission time (t_x), message arrival time (t_a), and message reception time (t_r). These measures are shown in relation to a message transmission in Figure 8-7.

As in Chapter 7, we define some performance parameters in terms of the basic operational quantities. These are summarized below:

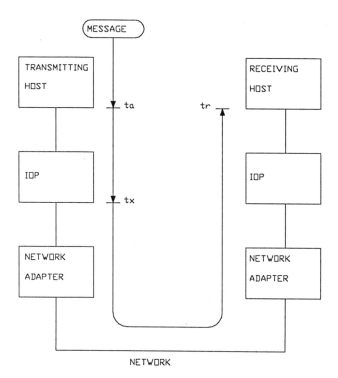

Figure 8-7 Message time tags.

190   Modeling and Analysis of Local Area Networks

Mean queue length	$N = W/T$	(8-9)
Mean response time	$R = W/C$	(8-10)
Utilization	$U = B/T$	(8-11)
Mean job service time	$S = B/C$	(8-12)
Network throughput	$X = C/T$	(8-13)
Network service time	$S_n = \sum_{\substack{\text{all}\\\text{messages}}} t_n/C \ ; \ \text{where } t_n = t_r - t_x$	(8-14)

The first five quantities are standard operational analysis results. The last, Equation (8-14), relates to the performance of the transmission mechanisms, ignoring the queue wait time.

NETWORK PERFORMANCE TESTS

An analysis run is performed on the test bed by initializing all network hosts with a known arrival rate generator and then by using the generated message traffic to load the network while collecting the operational measures defined above. After the run, the measures are combined and the desired performance measures are calculated.

The example test performed on the network test bed was formulated to give an indication of when, for a certain network configuration, the network becomes saturated (i.e., the network utilization approaches 1). For the example shown here, packet lengths of 200 and 400 bytes were tested with three nodes generating network traffic. The arrival rates at all three nodes were set up to be equal, and this rate varied from approximately 600 packets per second to approximately 15,000 packets per second. A test run was made for each of several arrival rates in the interval.

The mean queue length of packets awaiting transmission over the network for various arrival rates is shown in Figure 8-8. The values for each run using the measured values for $T(i)$ and the queue lengths at each arrival time are found through a combination of Equations (8-6), (8-7), and (8-9). Similarly, the mean response time was calculated using Equations (8-5), (8-7), and (8-11) and is plotted in Figure 8-9.

The utilization curve, shown in Figure 8-10, illustrates the percentage of the available data bandwidth that is being used to transmit message packets. From this graph, it can be seen that the particular network that we are analyzing approaches saturation (i.e., 100 percent utilization) rather quickly for the arrival rates and packet sizes shown.

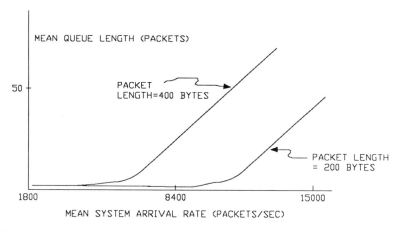

Figure 8-8 Mean queue length.

Figure 8-11 shows the effect of an increased arrival rate on service time. In this case, we have defined service time as the system busy time per completion, where the busy time considers the time a packet spends in the queue as well as the time it spends in transmission. When the system is saturated, however, there is always a packet ready to transmit, and so the queue fall-through time is "hidden" by this fact. Figure 8-12 illustrates this effect, which is known as "pipelining."

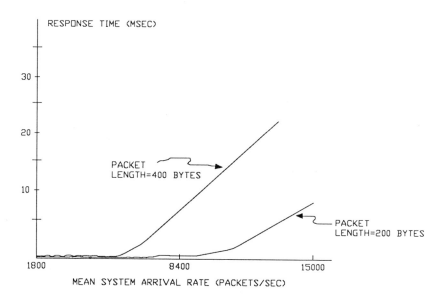

Figure 8-9 Mean response time.

192 Modeling and Analysis of Local Area Networks

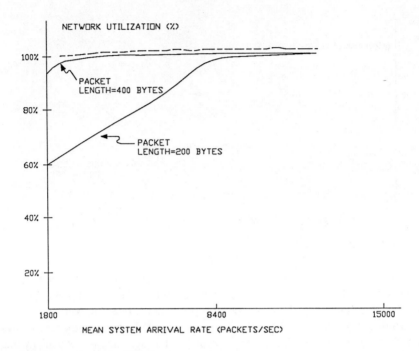

Figure 8-10 Network utilization.

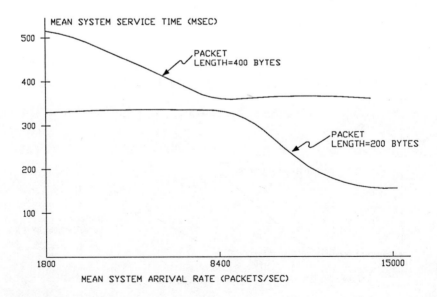

Figure 8-11 Mean system service time.

Hardware Test Beds 193

TIME	QUEUE POSITIONS			TRANSMISSION
1	PACKET 1			
2	PACKET 2	PACKET 1		
3	PACKET 3	PACKET 2	PACKET 1	
4	PACKET 4	PACKET 3	PACKET 2	PACKET 1
5	.	.	.	.

Figure 8-12 Pipeline effect on queue fall-through time.

In Figure 8-13, the network throughput is plotted against the system arrival rate. The results show that after saturation, network throughput for this type of network remains constant. This is an important property for some systems, especially since some network protocols cause degraded throughput under increased system load.

The final graph, in Figure 8-14, shows the time for a message to propagate through the network. This time does not include the queue wait time [see Equation (8-14)]. When the system is lightly loaded, this time will include the time for the token to travel around the network. Under heavy load, the time includes the delay associated with transmissions of message packets at other nodes.

This example has served a dual purpose: to illustrate the usefulness of hardware modeling in certain cases and to show the application of some of the operational analysis techniques discussed in Chapter 7. For hardware modeling, assumptions

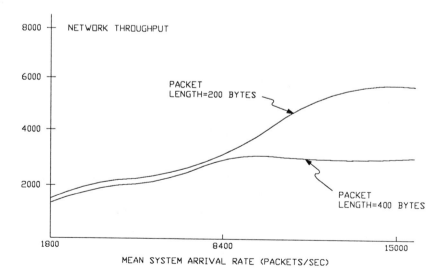

Figure 8-13 Network throughput.

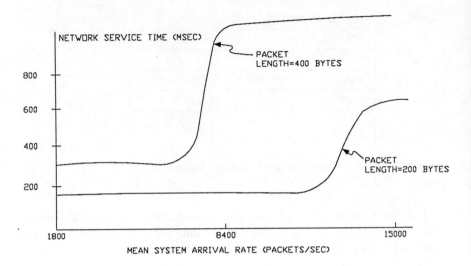

Figure 8-14 Network service time.

about the network behavior can be validated. Operational analysis enables us to calculate quantities of interest that either are not directly measurable or that are too difficult to measure without disturbing the actual operation of the network itself.

9

LAN Analysis

Chapter 4 introduced the basic concepts and theories embodied in analytical modeling. Addressed were basic concepts in queuing systems theory, its application to computer systems modeling, and an introduction to network modeling. This chapter will address the use of analytical models specifically from the viewpoint of use as an evaluation tool.

INTRODUCTION

In the past several years, the use of analytical performance models instead of the more widely used and familiar methods have become increasingly popular because of their relative simplicity of implementation and robustness of applications. These analytic models have been successful in estimation of such performance measures as throughputs, average queue lengths, and mean response times for real systems. This chapter is an introduction to queuing techniques for the modeling of computer communication networks, not an in-depth study.

The use of modeling to describe and imitate a real system has been with us since the beginning of the information revolution. These models are used not only to measure the performance of existing systems but also as part of the design and development of new systems. This latter goal is best attained through the use of analytical queuing models, as we will see in the following discussion of methods of performance evaluation.

The major performance evaluation tools (see Figure 9-1) other than queuing models are rules of thumb, linear projection, simulation, and benchmarking. These methods are listed in order of increasing complexity and implementation difficulty. The rules of thumb have been defined by the observation of operational systems and can be generally applied to local systems and extrapolated to distributed systems and networks. These rules take the following form:

196 Modeling and Analysis of Local Area Networks

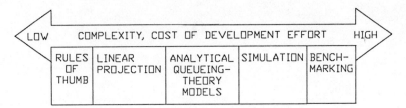

Figure 9-1(a) Spectrum of computer system modeling techniques.

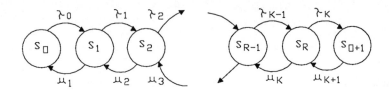

Figure 9-1(b) Spectrum of computer system modeling techniques.

1. Generally, channel use in direct access storage devices (DASD) should not exceed 35 percent for online and 40 percent for batch applications.
2. Individual DASD device use should not exceed 35 percent.
3. Average arm seek time on a DASD device should not exceed 50 cylinders.
4. No block size for auxiliary storage should exceed 4K bytes.

These rules are useful in that they are easy to apply, economical to use, and can be applied to day-to-day operations. They are limited in the sense that they cannot be used to predict the usefulness of hardware or software upgrades.

The linear projection method has been used to pick up on the rules of thumb at the prediction limitation point. Although results can be obtained, the accuracy of the results are limited by the fact that a linear projection is used to predict the behavior of inherently nonlinear systems. This method also requires the availability of an existing system to measure the pertinent performance criteria to be used as a base for the projection and estimation of future resource requirements.

For simulation and benchmarking, there is no absolute distinction between development and implementation costs. Simulation allows the model to contain much more detail than the other methods, but this may not be an advantage when compared to queuing methods where it has been found that too much information just serves to cloud the issue. Some simulation models are as large and cumbersome as the system they are modeling. The benchmarking method is the oldest and most used, but it is usually only helpful in the selection of the best hardware to process a known load. This is to say that the method requires existing hardware and, therefore, is not useful in the evaluation of hardware updates.

With the previous comments on other existing performance evaluation tools, we can assess the placement of queuing models and their overall usefulness. Queuing models reside between linear projection and simulation in terms of cost and complexity of implementation. Queuing models may be much simpler than the system they are modeling because only the most pertinent performance parameters need to be accounted for. Not only do queuing models have a place in the evaluation of existing systems, but they also may be used in the design and development phase of new systems to help in the selection of hardware and hardware-software interaction to avoid system bottlenecks.

According to [Trivedi, 1978], recent advances in modeling techniques are making analytical models increasingly capable of representing more and more aspects of the modeled system. Consequently, these techniques have been growing in popularity.

One method commonly used in system design is queuing analysis. Queuing models are more precise than other analytical techniques that predict performance based on average values [Nutt, 1978]. One reason is that queuing models allow greater detail to be used in describing systems, and hence they capture the more important features of the system. Oftentimes, several submodels are required; they are:

1. Workload model. Specifies the characteristics of the resource demands on various equipments in the system.
2. Configuration or system structure model. Specifies the hardware characteristics of the system.
3. Scheduling model. Specifies the scheduling algorithms whereby resources are allocated.

Queuing models can be categorized as either deterministic or stochastic in nature. If the design parameters to the model are known from prior experience or measurements, a deterministic analysis of the system may be carried out. Conversely, if the design parameters are not known, a stochastic analysis using various probability distributions is normally required.

Typical design parameters would include such items as:

1. The interarrival rate of events
2. Service times of these events
3. Number of servers being modeled
4. System capacity (i.e., number of events currently being processed and in queues)
5. Queuing discipline employed (i.e., FIFO, LIFO, etc.).

Normally queuing models provide some of the following performance attributes:

1. Average queue lengths
2. Average waiting time in queues
3. Use statistics
4. Average response times

Although queuing models have one overriding advantage in that they are cheap to use, there are a number of significant limitations to this method; they include:

1. Because these models assume the system has reached a steady state or equilibrium, peak or transient conditions are not modeled.
2. These models are limited as to the complexity of the problems that can be solved. As problems become more complex or additional details required, other methods must be used to model the systems.
3. Without actually measuring various design parameters, it is difficult to determine whether the characteristics of the data used will represent the system under investigation.

ANALYTICAL MODELING EXAMPLES

To better understand how these aforementioned techniques can be used to model and analyze a system, we will undertake two studies, one for the well-known Honeywell Experimental Distributed Processing system (HXDP) and the other for a lesser-known token bus. In both cases, similar quantities are sought; namely, average scan time (time for control to sequence around once) versus message size. The intent is to analyze the efficiency of the control protocol and network characteristics.

HXDP MODEL

Introduction

The HXDP system consists of processors connected to interface units that are joined by a bit-serial global bus. Bus allocation is governed by the vector-driven proportional access mechanism. Prior to system initialization, the 256-bit vectors are set for each processor so that for each time slice one and only one interface unit has a 1 in its index. The number of different schedules (possible combinations of 1s and 0s) for a system containing N interface units is, therefore, theoretically equal to $N \times N^{256}$, which is exorbitant even when $N = 2$. This scheme, however, cycles through the same pattern over and over again. Nonetheless, rather than develop a model allowing for any of the possible schedules, it was decided to constrain the allowable schedules to ease the computation.

It is assumed, consequently, that the schedule mechanism is as follows: every interface unit is assigned a 1 only once in the index, after which the initial pattern repeats itself until the 256th index for processor N is set; this pattern is termed a "scan block." The interface units, in turn, are sequentially logically numbered and are given a logical unit number.

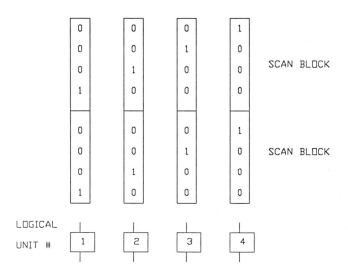

Figure 9-2 Scan blocks.

One illustration of such a schedule with the corresponding scan blocks would be for 4 IUs (see Figure 9-2). One schedule would contain $256/4 = 64$ scan blocks. Another schedule may be as shown in Figure 9-3. The sequence of events in the constrained system is then as follows:

The reallocation signal arrives at logical unit 1 (the IU with the first 1 bit); if there is a message waiting service, the interface unit is granted the bus and the

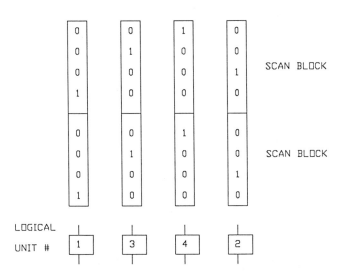

Figure 9-3 Another scan block schedule.

message is serviced. Once the message arrives at the destination, an acknowledgment is sent back to the source, after which a reallocation signal is sent out, the index is updated, and the next logical IU gets bus access. The sequence proceeds until IU N is serviced, after which (because of the assumptions) logical unit 1 is serviced. This continues ad infinitum.

The scan time is then the time it takes to scan through the logical sequence of IUs once; that is, through the scan block. Table 9-1 lists the terms and their definitions.

Analytical Modeling of the HXDP Bus

It is assumed that the messages are arriving at an exponential rate. The probability that an arbitrary interface unit will require service during one scan is

$$P = \int_{-\infty}^{\infty} (1 - e^{-\lambda t}) f_\tau(t) dt$$

where $f_\tau(t)$ is the probability density function of the scan time.

For λt small, we use the approximation that $e^{-\lambda t} \approx 1 - \lambda t$. Thus

$$P = \int_{-\infty}^{\infty} \lambda t f_\tau(t) dt = \lambda \overline{\tau}$$

Table 9-1 Symbols and Their Definitions

Symbols	Definitions
N	Number of interface units in the system
T_{si}	Time requirement to service a message at interface unit i
T_{ri}	Time delay associated with the reallocation signal passing from interface unit with logical sequence i to i + 1
λ_v	Average message arrival rate at interface unit i
$\underline{\tau}$	Time to scan through entire sequence of IUs
τ	Average or expected time scan
∂_i	Set equal to 1 or 0 depending upon whether IU has message awaiting transmittal or not
d	Distance between physical interface units i and i + 1
T_s	The time it takes to send the message of predetermined constant size from IU i to IU i + 1 separated by distance d
t_{ack}	The time it takes to send the acknowledgement from IU i to IU i + 1 separated by distance d.
t_R	The time it takes to send the reallocation signal from IU i to IU i+1 separated by distance d
p	The probability of an arbitrary interface unit requiring service during one scan

For varying message arrival rates $= \lambda_i$, this evaluation becomes

$$P_i = \int_{-\infty}^{\infty} \lambda_i t f_\tau(t) dt = \lambda_i \bar{\tau}$$

$\nabla \tau_i$ which is the segment of the scan time that can be attributed to IU i is

$$\nabla \tau_i = \zeta_i (T_{si} + T_{ack}) + T_r ;$$

where T_{ACK} is the time it takes the for acknowledgement (ACK) to be sent back to i, if there was a message received, and T_R is the time it takes for the reallocation signal to go from IU logical number i to IU logical i + 1. But ζ-i can only take on values 0 or 1, and since we assume a uniform destination distribution and an IU in the HXDP system communicates with itself via the bus

$$E(\zeta_i \cdot (T_{si} + T_{ack})) = \frac{1}{n} \sum_{k=1}^{N} (|i - k| \cdot (t_s + t_{ack})) \cdot P_i$$

This implies that

$$\tau + \frac{1}{N} \sum_{i=1}^{N} \sum_{k=1}^{N} |i - k| \cdot (t_s + T_{ack}) P_i + \sum_{i=1}^{N} |i - (i+1)| t_R$$

where |i-k| represents the number of interface units away from the source interface unit i that interface unit k is located.

Substituting $\lambda_i \bar{\tau} \sim P_i$ we get

$$\bar{\tau} = \frac{\bar{\tau}(t_s + t_{ack})}{N} \sum_{i=1}^{N} (i^2 + (1+N)(-i + \frac{1}{2}N)) \cdot \lambda_i + \sum_{i=1}^{N} |i - (i+1)| t_R$$

Thus,

$$\bar{\tau} = \frac{\sum_{i=1}^{N} |i - (i+1)| t_R}{1 - \frac{t_s + t_{ack}}{N} (\sum_{k=1}^{N} (k^2 + (1+N)(-k + \frac{1}{2}N)) \lambda_k)}$$

where index i denotes the logical number of the interface unit.

The main constraint on the model is

$$\frac{t_s + t_{ack}}{N}\left(\sum_{k=1}^{N}(k^2 + (1+N)(-K+\frac{1}{2}N)\right) \cdot \lambda_K) \ll 1$$

The effect on the average scan time, $\overline{\tau}$, can now be determined by varying any combination of the following parameters:

1. The number of interface units, N
2. The arrival rates, λ_k
3. The logical numbering of the interface units
4. The distances between neighboring interface units
5. The average size of the arriving messages

GRAPHIC OUTPUTS

Figures 9-4 through 9-7 show the resultant computations for the scan time versus the message size for various changes in arrival rate, processor location, and quantity. These results will be compared with those of Malan later in this chapter.

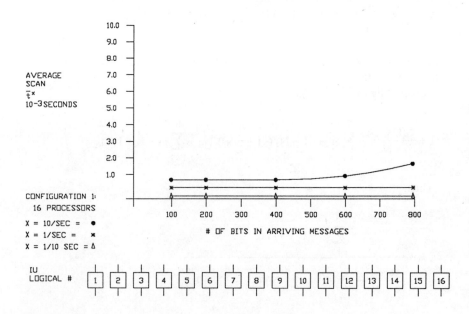

Figure 9-4 Scan time versus message size configuration 1.

LAN Analysis 203

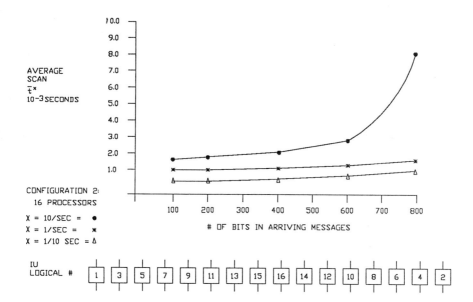

Figure 9-5 Scan time versus message size configuration 2.

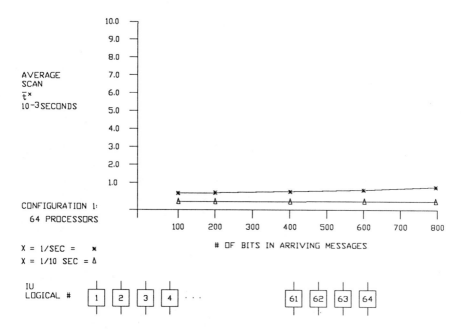

Figure 9-6 Scan time versus message size configuration 1b.

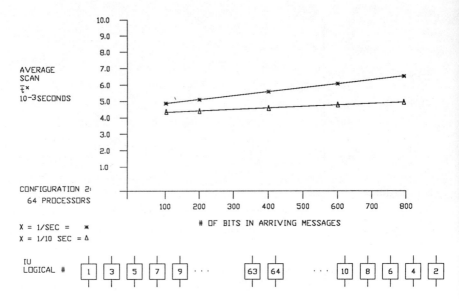

Figure 9-7 Scan time versus message size configuration 2b.

TOKEN BUS DISTRIBUTED SYSTEM

Introduction

The token bus distributed processing system (a local computer network) consists of processors connected to interface units which, in turn, are connected by a common communication medium—the global bus. The allocation of the bus is controlled by the cyclic passing of tokens in a sequential manner from lowest numbered interface unit (IU) to the next highest until all numbered IUs have been interrogated and serviced. The sequential numbering is determined during power-up and once steady state has been reached may be assumed to remain constant for modeling purposes. If an IU requires no service, control is passed to the next IU with an associated delay. The time it takes for the control to pass through the sequence completely is termed the scan time (as previously discussed).

Rather than investigating the entire token bus system, per se, emphasis will be on the bus, or the IU and bus layer, for modeling purposes. The main body of the example documents the development of the analytical models, in particular, the solution of the models for the value of the average scan time. The analytic computation of the scan time allows one to further determine such interesting and practical bus parameters as average message waiting time, average queue length, and bus use. From the derived formulas, one can readily ascertain the effect on bus parameters of increasing the number of processors, altering the sequential placement of processors, or varying the message arrival rates.

Preliminary Formulations and Definitions

The message arriving at the processor is assumed to follow a Poisson distribution; in other words,

$$P(R,t) = \frac{(\lambda t)^R e^{-\lambda t}}{R!} \quad (R = 0, 1, 2 \ldots)$$

where P[r,t] is the probability that r messages arrive in time t, with each message being of the same size.

Service is required if there are one or more message arrivals in time t or

$$P(1 \text{ or more arrivals, } t) = 1 - P(0, t), \text{ which, in turn, } = 1 - \frac{(\lambda t)^0}{0!} e^{-\lambda t} = 1 - e^{-\lambda t}$$

Since the assumption is that steady state has been reached, we can let $f_\lambda(t)$ denote the probability density function of the scan time $\tau, F_\lambda(t)$; the corresponding cumulative distribution is then equal to:

$$F_\tau(t) = P(\tau \leq t); \quad \frac{dF_\tau}{dt} = f_\tau(t); \quad -\infty < t < \infty$$

$$F_\tau(t) = \int_{j\infty}^{t} f_t(u) du$$

Let p denote the probability of an arbitrary processor requiring service during one scan,

$$P = \int_{-\infty}^{\infty} (1 - e^{-\lambda t}) P_\tau(t) \, dt$$

If more than one arrival occurs at any processor in any scan, that arrival can be considered to be blocked. This message will then have to wait at least one scan time before it can be placed on the bus. This implies that:

P (blocking) = P (more than one arrival in scan time)

P (more than one message requiring service in time t) = 1 − [P(0t) + P(1, t)]
$$1 - [e^{-\lambda \tau} + \lambda t e^{-\lambda t}]$$

To enable the evaluation of p, for λt small, $e^{-\lambda t} \approx 1^{-\lambda t}$, which implies that

$$P = \int_{-\infty}^{\infty} (1 - e^{-\lambda t}) f_\tau(t) \, dt = \int_{-\infty}^{\infty} \lambda t f_\tau(t) dt = \lambda \int_{-\infty}^{\infty} t f_\tau(t) dt$$

which is equal to $\lambda \tau$, by definition of expected value.

The relationship of $p \approx \lambda \bar{\tau}$ will be used for all the models for simplification purposes. For clarity and convenience, Table 9-2 contains the symbols and their corresponding definitions which will be used in the development of the analytical models.

Analytical Modeling of the Token Bus

The analytical models developed for the token bus will be presented in an order reflecting an increasing degree of complexity and, consequently, a relaxation of the corresponding mathematical assumptions. In each of the models, a steady state, constant message size, and equal spacing between processors will be assumed. In addition, once the IU has been given control of the bus, it will be assumed that the message buffer for the interface unit will be emptied instantaneously onto the bus. The underlying specific assumptions in each case will be clearly outlined.

Table 9-2 Symbols and Their Definitions

Symbols	Definitions
N	Number of processors and, consequently, number of interface units due to a one-to-one correspondence in the system.
T_s	Time required to service a message at an interface unit
T_{si}	Time required to service a message at an interface unit i.
T_c	Time delay associated with control (token) passing from an interface unit to its physically nearest neighbor interface unit
T_{ci}	Time delay associated with control passing from interface unit with logical sequence number i to interface unit with logical sequence number (i + 1)
λ	Average message arrival rate at each interface unit
λ_i	Average message arrival rate at interface unit i
$\underline{\tau}$	Time to scan through the entire sequence of IUs
τ	Average or expected scan time
∂_i	Set equal to 1 or 0 depending upon whether interface unit i has a message awaiting transmittal or not
d	Distance between interface units i and i +1
ts	The time it takes to send the message of predetermined constant size from IU i to IU i + 1 separated by distance d
UF	Bus use factor
p	Probability of an arbitrary interface unit requiring service during one scan

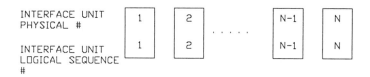

Figure 9-8 Mappimg of physical IDs to logical ones.

Case 1

In the basic analytical model, it will be assumed that the arrival rate of messages at each of the N interface units is equivalent and is represented by λ. In addition, it is assumed that once steady state has been reached, the sequential (logical) numbering of the interface units is identical to the physical numbering (spatial numbering from left to right); that is, it follows the representation shown in Figure 9-8.

If we let Tc denote the time delay associated with the token (control) passing from one interface unit to another, the Tci for each interface unit may be considered the same since it has been assumed that the processors are equidistant from one another, and, consequently, the control will need to traverse the same distance from a processor to its next (with next highest sequence number) neighbor.

Another essential time parameter is the time it takes to service a message for any interface unit. In a ring topology with a token passing scheme one could consider Ts, which is the time required to service a processor, to average out to the same value over time for all processors. Yuens [1972] shows that the same conclusion cannot be reached for the bus topology. Time to service a message is a function to the destination IU. Therefore, the placement of the source IU within the bus topology will affect the average time it takes to service one of its messages. For example, if N = 3, we have the configuration shown in Figure 9-9.

For interface unit 1 to transmit a message to processor 2, the message will have to traverse the distance from 1 to 2; for interface unit 1 to transmit a message to 3, it will have to traverse the distance from 1 to 3. If we represent the equal distance between two neighboring interface units as d and we let each of the other IUs be potential similar message destinations (that is, a uniform distribution for message destinations is assumed),

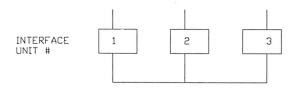

Figure 9-9 Token bus configuration.

208 Modeling and Analysis of Local Area Networks

$$E(T_s/\text{source} = 1) = \frac{1}{2} \cdot \left(\frac{d}{\text{velocity estimate}}\right) + \frac{1}{2}\left(\frac{2d}{\text{velocity estimate}}\right)$$

where $\dfrac{d}{\text{velocity estimate}} = t_s =$ time to send the message of the chosen size from i to $i+1$.

$$= \frac{1}{2}t_s + t_s = \frac{3}{2}t_s$$

For processor 2 as the source processor, the corresponding equation becomes

$$E(T_s/\text{source} = 2) = \frac{1}{2}t_s + \frac{1}{2}t_s = t_s$$

Let Ts_i denote the time it takes to service a message at interface unit i. The time it takes for the token to pass from the ith IU to the $i+1$st interface unit may then be expressed as

$$\nabla \tau_i = \zeta_i T_{si} + T_c; \quad \zeta_i = \begin{cases} 1, & \text{if IU; has a message awaiting transmittal} \\ 0, & \text{otherwise} \end{cases}$$

The total scan time becomes

$$(1) \quad \tau = \sum_{i=1}^{N} \nabla \tau_i \text{ or } \tau = \sum_{i=1}^{N} (\zeta_i T_{si} + T_c)$$

Now, taking expectations of both sides of Equation (1) we get

$$\text{average value of scan time} = \overline{\tau} = \sum_{i=1}^{N} E(\zeta_i T_{si}) + \sum_{i=1}^{N} E(T_C) = \sum_{i=1}^{N} E(\zeta_i T_{si}) + NT_C$$

Next, by definition of the expected value of product,

$$E(\zeta_i T_{si}) = \sum_{\zeta_i} \cdot \sum_{T_{si}} \zeta_i \cdot T_{si} \cdot P(\zeta_i \cdot T_{si})$$

Since it was mentioned previously that Tsi is a function of the distance that the message has to travel and since ζ_i can only take on the value 0 to 1,

$$E(\sigma_i T_{si}) = \frac{1}{N-1} \sum_{k=1}^{N} |i - k| \cdot t_s \cdot P$$

where i-k represents the number of interface units away from the source interface unit i, the destination interface unit k is located and any interface unit other than i has an equally likely probability of being a destination interface unit. That is, a probability $= \frac{1}{N-1}$.

"p" is the probability derived in the preliminary formulation.

In summary,

$$\bar{\tau} = \frac{1}{N-1} \sum_{i=1}^{N} \sum_{k=1}^{N} |i-k| \cdot t_s P + NT_C;$$

$$\bar{\tau} = \frac{t_s \cdot P}{N-1} \sum_{i=1}^{N} \sum_{k=1}^{N} |i-k| NT_C = \frac{t_s \cdot p}{N-1} \left(\frac{N^3 - N}{3} \right) + NT_C$$

$$= t_s \cdot P \cdot \frac{N(N+1)}{3} + NT_C$$

Substituting $P \approx \lambda \bar{\tau}$ into the above equation,

$$\tau = \frac{t_s \cdot \lambda \bar{\tau} N (N+1)}{3} + NT_C$$

$$\bar{\tau}(1 - \frac{t_s \cdot \lambda \cdot N (N+1)}{3}) + NT_C$$

(2) $\bar{\tau} = \frac{NT_C}{1 - \lambda N(N+1) t_s/3}$

This equation is valid, based upon the assumption and approximations if and only if

$$\frac{\lambda N(N+1) t_s}{3} << 1$$

Case II

In this model, we relax the assumptions that all the interface units have identical message arrival rates equal to λ by allowing for message arrival rates of λ for interface unit i. However, we retain the assumption of the hypothetical logical or physical configuration which, in turn, will be eliminated in the subsequent case. The relaxation of the assumptions is being done in a gradual manner to emphasize the evolutionary nature of the development of the analytical models. The relaxation of the equivalent message arrival rates will allow for a greater realm of applicability and consequently of testing but will, naturally, complicate the ultimate formula for τ.

Since each interface unit now has a characteristic message arrival rate λ, for interface unit i, p now becomes

$$P_i = \int_{-\infty}^{\infty} (1 - e^{-\lambda_i t}) f\tau(t) dt$$

$$P_i \approx \lambda_i \bar{\tau}$$

Equation (1) is still applicable, that is,

$$\tau = \sum_{i=1}^{N} \nabla \tau_i = \sum_{i=1}^{N} (\zeta_i T_{Si} + T_C)$$

Furthermore, $\bar{\tau}$ is still

$$\tau = \sum_{i=1}^{N} E(\zeta_i T_{Si}) + N T_C$$

where

$$E(\zeta_i T_{Si}) = \frac{1}{N-1} \sum_{k=1}^{N} |i-k| \cdot t_s \cdot P_i$$

$$\bar{\tau} = \sum_{i=1}^{N} \frac{1}{N-1} \sum_{k=1}^{N} |i-k| \cdot t_s \cdot P_i + NT_C$$

$$\bar{\tau} = \frac{t_s}{N-1} \sum_{i=1}^{N} (i^2 + (1+N)(-i+\frac{1}{2}N)) \cdot P_i + NT_C$$

Substituting $\lambda_i \bar{\tau} \approx P_i$,

$$\bar{\tau} = \frac{\bar{\tau} t_s}{N-1} \sum_{i=1}^{N} (i^2 + (1+N)(-i+\frac{1}{2}N)) \cdot \lambda_i + NT_C$$

Thus, Equation (3) is

$$\bar{\tau} = \frac{NT_c}{1 - \frac{t_s}{N-1} (\sum_{i=1}^{N} (i^2 + (1+N)(-i+\frac{1}{2}N)) \cdot \lambda_i)}$$

Case III

This model incorporates major modifications which should permit the model to better reflect the actual system. In particular, it is assumed that once steady state has been reached, the logical numbering does not have to reflect the physical location, but in fact, the steady state configuration may be, for example, as shown in Figure 9-10.

The logical sequence numbering of the token bus system may assume any out of the N! possible different choices of the steady state with an equal probability. Therefore, it is of the utmost importance to develop a model which can reflect all N! of the possible combinations.

Consequently, T_C, now, is not a constant but must in some sense reflect the time it takes for the token to travel the distance from interface unit i to interface unit i + 1.

Let i represent the logical sequence number of the interface unit; then $\nabla\tau_i = \zeta_i T_{si} + T_{ci}$; where T_{ci} = time for token to traverse the distance from the IU with logical sequence i to the IU with logical sequence i + 1 (N + 1 - st IU becomes IU 1).

In the previous models, the logical sequence number and the physical number of the interface units were identical; therefore, it was not necessary to state explicitly the correspondence of the index i.

The total scan time is now expressed by

$$(4) \quad \tau = \sum_{i=1}^{N} \nabla\tau_i \quad \text{or} \quad \tau = \sum_{i=1}^{N} \zeta_i T_{si} + T_{Ci}$$

The average value of scan time equals

$$\tau = \sum_{i=1}^{N} E(\zeta_i T_{si}) + \sum_{i=1}^{N} E(T_{Ci})$$

which for a known configuration is equal to

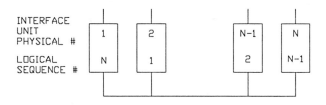

Figure 9-10 Mapping logical to physical.

$$\sum_{i=1}^{N} E(\zeta_i T_{si}) + \sum_{i=1}^{N} T_{Ci} \quad \text{or} \quad \sum_{i=1}^{N} E(\zeta_i T_{si}) + \sum_{i=1}^{N} |i - (i+1)| t_c$$

(N + 1 denotes 1, due to cycling), where t_c is the time it takes for the control to pass from any interface unit k to physical unit k+1.

We have previously evaluated

$$\sum_{i=1}^{N} E(\zeta_i T_{si}),$$

and in order to take advantage of the results, the index k will denote the physical number of the interface unit, while the index i will denote the logical number of it.

Combining the results and making the substitution

$$\lambda_k \bar{\tau} \approx P_k,$$

we get

(6) $$\bar{\tau} = \frac{\sum_{i=1}^{N} |i - (i+1)| t_c}{1 - \frac{t_s}{N-1} (\sum_{k=1}^{N} (k^2 + (1+N)(-k + \frac{1}{2}N)) \cdot \lambda^k}$$

From Equation (6), the effect on the average scan time can be determined by varying any combination of the following variables:

1. The number of interface units, N
2. The arrival rates, λk, of the messages at the interface units with logical numbers, k.
3. Varying the logical sequential numbering of the interface units
4. Varying the distances between neighboring interface units
5. Varying the average size of the messages arriving at the interface units

The main constraint of the model is

$$\frac{t_s}{N-1} (\sum_{k=1}^{N} (k^2 + (1+N)(-k + \frac{1}{2}N)) \cdot \lambda^k) \ll 1$$

LAN Analysis

Derivation of the expected number of messages in queue and the expected waiting time is

The average number of messages in the token bus distributed processing system, $\overline{M}$, for the IU that are either being serviced by the bus or are waiting in the queue for service for an average scan time $\overline{\tau}$ and average message arrival rate λ, is equal to

$$\text{average number of messages for IU} = \overline{M} = \sum_{i=0}^{\infty} i \cdot P(\text{number of messages arriving} = i)$$

$$\sum_{i=0}^{\infty} i \frac{(\lambda\overline{\tau})^i e^{-\lambda\overline{\tau}}}{i!} = (\lambda\overline{\tau}) e^{-\lambda\overline{\tau}} \sum_{i=0}^{\infty} \frac{(\lambda\overline{\tau})^{i-1}}{(i-1)!}$$

(7) $\overline{M} = (\lambda\overline{\tau})$

$$= e^{-\lambda\overline{\tau}} \cdot (\lambda\overline{\tau}) \cdot \sum_{j=0}^{\infty} \frac{(\lambda\overline{\tau})^J}{J!}$$

The average queue size at the IU is then

$$\text{average number of messages in queue for IU} = \overline{M}_Q = \sum_{i=0}^{\infty} (i-1) \cdot \frac{(\lambda\overline{\tau})^i e^{-\lambda\overline{\tau}}}{i!}$$

$$\overline{M}_Q = \sum_{i=1}^{\infty} \frac{i(\lambda\overline{\tau})^i e^{-\lambda\overline{\tau}}}{i!} - \sum_{i=1}^{\infty} \frac{i(\lambda\overline{\tau})^i e^{-\lambda\overline{\tau}}}{i!}$$

$$\overline{M}_Q = \overline{M} - \sum_{i=1}^{\infty} \frac{(\lambda\overline{\tau})^i e^{-\lambda\overline{\tau}}}{i!} \qquad \overline{M}_Q = (\lambda\overline{\tau}) - (1 - e^{-\lambda\overline{\tau}}) = (\lambda\overline{\tau} + e^{-\lambda\overline{\tau}}) - 1$$

The total number of messages in the system for N processors, all having message arrival rates $= \lambda$, is then

(9) $N \cdot \overline{M}$

The average number of messages in the system awaiting service is

(10) $N \cdot \overline{M}_Q$

If each interface unit has a different λ, we can let

$$\overline{\lambda} = \frac{1}{N} \sum_{i=1}^{N} \lambda_i$$

Equation (7) then becomes

$M = \lambda \bar{\tau}$

Equation (8) becomes

$M_Q = (\lambda \bar{\tau}) + (e^{-\lambda \bar{\tau}}) - 1$

Equation (9) and (10) hold.
To evaluate the expected waiting time for a message in queue at each IU:

E (waiting time) = E (number in queue) x $\bar{\tau}$

or

$M_Q \cdot \bar{\tau} = \lambda \bar{\tau}^2 + \bar{\tau}(e^{-\lambda \bar{\tau}} - 1)$

Another useful measurement is the probability that the system is blocked. This is equivalent to the probability that more than one message arrival occurs at an interface unit in any scan time. Letting λ denote the message arrival rate at any interface unit,

P(blocking) = P(more than 1 message arrival in scan time)
$= 1 - (e^{-\lambda \bar{\tau}} + \lambda \bar{\tau} e^{-\lambda \bar{\tau}})$
$= 1 - e^{-\lambda \bar{\tau}} (1 + \lambda \bar{\tau})$

The fraction of message being blocked equals

$\dfrac{\text{P(more than 1 messagee arrivall in scan time)}}{\text{P(terminalll requiress servicee)}}$

$= \dfrac{1 - e^{-\lambda \bar{\tau}} (1 + \lambda \bar{\tau})}{1 - e^{-\lambda \bar{\tau}}}$

Derivation of Token Bus Use

Another practical measurement which may be determined given the previous analytical results is the bus use factor. This quantifies the use of the bus for actual message transmission.

Utilization factor = $\dfrac{\bar{\tau} - NT_C}{\bar{\tau}}$ for a first approximation, as in Case I.

$$\text{or } \frac{\overline{\tau} - \sum_{i=1}^{N} |i - (i+1)| t_c}{\tau} \text{ where } \tau \text{ as in Case III.}$$

This equation demonstrates that as the distance between sequential neighbors increases, the actual bus use decreases because more time is spent in token passing. To obtain an even more precise factor, subtract from the numerator the estimated time allocated for transmission of the overhead bits per message.

Graphic Outputs

The results of these computations for scan rate versus message bits as the arrival rate and configuration changes are shown in Figure 9-11 through 9-14.

SUMMARY

This chapter illustrated the usefulness of analytical modeling for studying a component of a local area network, in this case the control scan time. These models provide a fairly easy means to extract such information. They have been applied to a variety of other LAN problems. The reader interested in more details of such modeling is directed to acquire the references cited in this chapter.

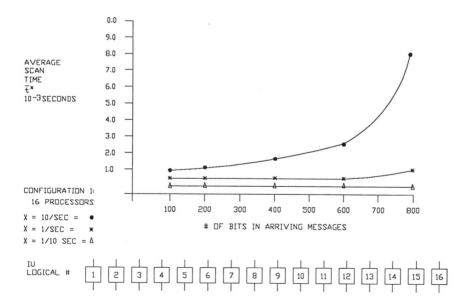

Figure 9-11 Token bus scan rate C1.

216 Modeling and Analysis of Local Area Networks

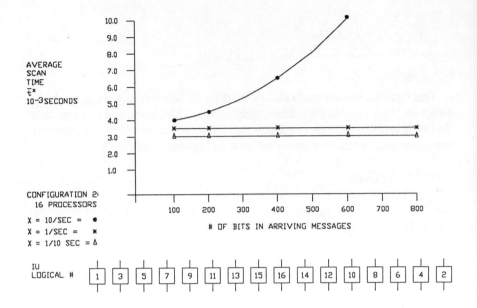

Figure 9-12 Token bus scan rate C2.

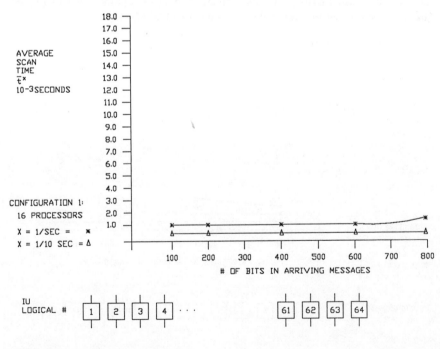

Figure 9-13 Token bus scan rate C1'.

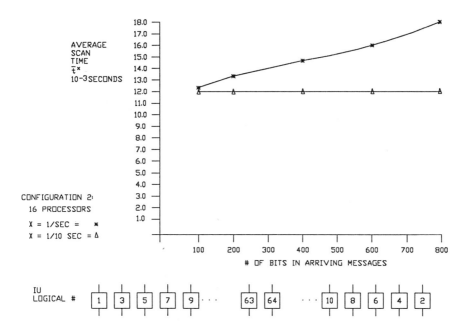

Figure 9-14 Token bus scan rate C2'.

REFERENCES

Kleinrock, L. *Queuing Systems Vol I and II*, John Wiley and Sons, 1975 and 1976.

Little, J. D. *A Proof of the Queueing Formula: L = λW*, Operations Res. 9(3), 1961.

Nutt, G.: "A Case Study of Simulation as a Computer Design Tool," IEEE Computer, October 1978, pp. 31-36.

Trivedi, K.: "Analytic Modeling of Computer Systems," Computer, October 1978, pp. 38-56.

Yuen, M. "Traffic Flow in a Distributed Loop Switching System," in Proceedings of Symposium on Computer Communications Networks and Teletraffic, Polytechnic Institute of Brooklyn, April 4-6, 1972.

10

MALAN

INTRODUCTION

Selecting the most cost-effective local area network (LAN) design from an overwhelming number of options is a complicated and time-consuming affair. The selection process consists of a wide array of functions such as:

- Network assessment (Do we need a LAN; is the present LAN effective?)
- Performance assessment (interface units, protocols, and links).
- Network optimization (Configurations and protocols).
- Management support assessment.

All these functions and many others have been addressed in the previous chapters in terms of design and selection methodology and as a potential means to aid in the process. This chapter will continue this discussion by providing a framework upon which models of a limitless arrangement of LAN architectures and their features can be developed and used to aid in this process.

SIMULATING LOCAL AREA NETWORKS

Developing a simulation general enough to model a wide range of local area network architectures requires an up- front understanding and definition of the components that comprise all networks, as well as a characterization of the interaction of the various components. We cannot hope to develop a tool to analyze networks without having a general view and understanding of the components that make up any local area network. The following section generalizes the components of a LAN and highlights their essential pieces.

COMPUTER NETWORKS (THE MODEL)

A computer network can be considered to be any interconnection of an assembly of computing elements (systems, terminals, etc.) together with communications facilities that provide intra and internetwork communications.

These networks range in organization from two processors sharing a memory to large numbers of relatively independent computers connected over geographically long distances. (The computing elements themselves may be networks, in which case it is possible to have recursive systems of networks ad infinitum.) The basic attributes of a network that distinguish its architecture include its topology or overall organization, composition, size, channel type and utilization strategy, and control mechanism.

Using the nomenclature and taxonomy of Anderson [1975] for computer interconnection structures, a particular system can be characterized by its transfer strategy (direct or indirect), transfer control mechanism (centralized or decentralized), and its transfer path structure (dedicated or shared). Various network topologies, such as ring, bus, and star, are seen as embodiments of unique combinations of the above characteristics (see Figure 10-1).

Network composition can be either heterogeneous or homogeneous, depending on either the similarity of the nodes or the attached computing elements. Network size generally refers to the number of nodes or computing elements. With respect to its communications channels, a network may be homogeneous or it may employ a variety of media. Overall network control or management is usually either highly centralized or completely distributed. If the hardware used for passing line control

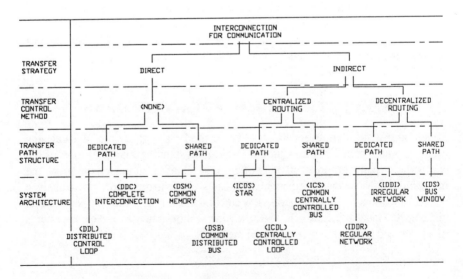

Figure 10-1 Taxonomy of computer interconnection structures.

from one device to another is largely concentrated in one location, it is referred to as "centralized control." The location of the hardware could be within one of the devices that is connected to the network, or it could be a separate hardware unit. If the control logic is largely distributed throughout the different devices connected to the network, it is called "decentralized control."

Implementation-independent issues that are dependent on the above system attributes are modularity, connection flexibility, failure effect, failure reconfiguration, bottleneck, and logical complexity. These issues are explored in Anderson [1975]. A subset of all possible computer systems is that of local computer networks (LCNs). Although no standard definition of the term exists, an LCN is generally regarded as being a network so structured as to combine the resource sharing of remote networking and the parallelism of multiprocessing. A usually valid criterion for establishing a network as an LCN is that its internodal distances are in the range of 0.1 to 10 km with a transfer rate of 1 to 10 mbps.

Bus-Structured LCN

The range of systems to be initially studied will be confined to what is known in the LCN taxonomy of Thurber [1974] (see Figure 10-2) as category 3 bus structured systems. As opposed to such point-to-point media technologies as circuit and message switching, a bus-structured system consists of a set of shared lines that can be used by only one unit at a time. This implies the need for bus-control schemes to avoid inevitable bus usage conflicts.

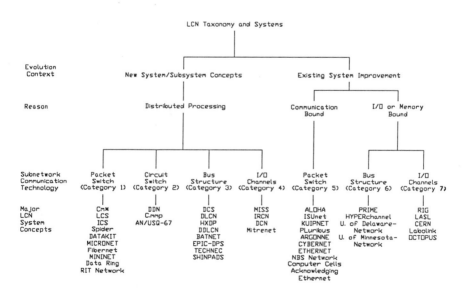

Figure 10-2 Local computer network taxonomy.

Network Components

As a first step in developing a general LCN simulation, the network model illustrated in Figure 10-3 is established. A network consists of an arbitrary number of interconnected network nodes. Each node consists of one or more host computing elements or processors connected to an independent front-end processor termed an interface unit (IU).

The hosts are the producers and consumers of all messages, and they represent independent systems, terminals, gateways to other networks, and other such instances of computing elements. The IUs handle all nodal and network communication functions such as message handling, flow control, and system reconfiguration. The lines represent the physical transmission media that interconnect the nodes. The IUs together with the line interconnection structure comprise the communication subnetwork.

The model in Figure 10-3 isolates the major hardware units involved in the transfer of information between processes in different hosts. At this level no distinction is made between instances of messages such as data blocks and acknowledgments. In order to develop and refine the model, the major elements, structures, and activities must be further defined.

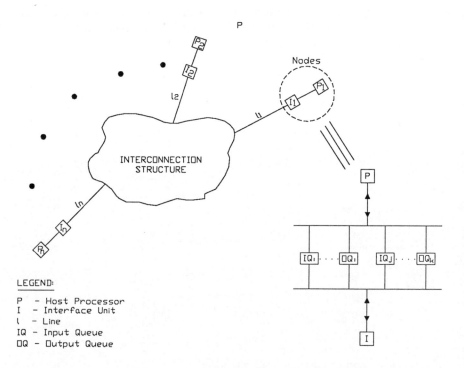

LEGEND:
P - Host Processor
I - Interface Unit
l - Line
IQ - Input Queue
OQ - Output Queue

Figure 10-3   General distributed computer network.

Host Processors

Host or processor components generally include computation and control elements, various levels of memory, and input and output peripherals. As far as the system is concerned, each processor's behavior can be considered to be reflected in appropriate distribution functions that describe the rate at which the processor produces and consumes interprocessor messages. These functions reflect a given processor's inherent processing power and loading based on processor parameters, exogenous communication levels, and internetwork communications.

Queues

Queues are shared memory buffer structures through which information transfer between a processor and its IU takes place. For each node there will be an output (line) queue for messages awaiting transmission as well as one or more input (message) queues containing unprocessed receptions. The queue memory area may be located in the processor or in the IU depending on the implementation. Functionally, both are equivalent.

Associated with queues are control variables that are maintained and monitored by both the processors and IUs to provide for the simultaneous and asynchronous access of the queues. The most common types are linear, circular, and linked queues. Linear queues (buffers) are used when the extent of a message is known and the buffer structure can be allocated in advance. The use of circular buffers is appropriate if several messages of undetermined length are to be buffered before one of them is processed. A pool of chained queues is used if the message sizes and arrival times vary over wide ranges that cannot be predicted in advance and the messages are not removed in order of their arrival.

Messages are deposited (written) into and withdrawn (read) from queues using various strategies such as FIFO (first-in, first-out), LIFO (last-in, first-out) and longest message first.

Queue access is controlled in order to prevent writing into a full queue, reading from an empty queue, and reading information as it is being written.

Interface Units

Insofar as their role in the network is concerned, the Interface unit is the most complex unit with respect to both hardware and software. The basic function of the IU is to enable its processor to communicate with others in the network as well as to contribute to overall network functioning. This involves system (re)initialization, flow control, error detection, and management.

When the IU detects that its processor has a message to send, it formats the message for transmission and becomes a contender for exclusive use of the communications channels. Upon allocation of control, the controller transmits the message and, depending on the implementation, may await a response from the destination processor.

Upon completion of resource (bus) utilization, the IU must be able to pass control to the next candidate according to the allocation scheme. If there is an IU failure, the other IUs must be able to substitute for it insofar as its network control responsibilities are concerned.

Communication Lines

The lines are the physical connections between network nodes over which control and data transmissions travel. Common equivalent terms are "channel" and "circuit." A particular circuit is either uni- or bidirectional (by nature and/or use) and supports continuous transmissions provided by analog or digital techniques.

Circuits are supported using a variety of media such as coaxial cables, twisted pairs, fiber optics, microwave links, laser links, etc. For the purposes of the simulation it is not necessary to be concerned about these low-level characteristics except as they are represented by a set of channel characteristics: the maximum data rate, delay and error parameters, and directional limitations.

It is also useful to consider setup characteristics if a point-to-point circuit is not always dedicated to a network. These setup characteristics may include the signaling mechanism and delay, circuit setup delay, and the delay for breaking the circuit. In the systems that will be examined later in the chapter, setup characteristics will not be a factor.

Maximum data rates vary from 50 kbps (twisted pair) up to 150 mbps (optical cables). This rate represents the raw transmission capability of the line and is not the same as the net rate at which information is transferred. There is always an overhead. Various factors such as logic failures, electronic interference, and physical damage give rise to transmission degradations ranging from single-bit errors to total line failure. Depending on the type of line used, typical error rates vary from 1 in 10,000 to 1 in 10,000,000 bits transmitted.

Interconnection Structures

Various network aspects such as scheduling, message routing, and reconfiguration are fundamentally related to a network's physical interconnection structure. For example, eligibility for bus control may be dependent on position, the time for a message transfer may be dependent on the location of the processors involved, or a network's continued functioning may be contingent upon the existence of a redundant link. This structure may be represented by a topological organization of the three hardware archetypes — nodes, paths and switches — that are involved in the transfer of information between processes at different nodes.

These transfers are called "message transmissions" and do not distinguish between instances of messages such as data blocks, service requests, semaphores, etc. Likewise, in restricting consideration to structural issues it is unnecessary to distinguish between a computing element and its interface. They are lumped together as the entity "node."

The switching elements are "intervening intelligences" between a source and a destination of a message that affect the routing or the destination in some way.

Figure 10-4 shows a general model of an interconnected system. For simplicity, the class of systems with only one switch is represented.

Associated with each node and switch is a number of paths or links. Each node can connect to the rest of the network through one, two, or multiple links corresponding, respectively, for example, to a bus system, a ring or loop structure, or a fully interconnected network with direct links between each pair of nodes. Figure 10-5 shows specific examples of interconnection structures.

These diagrams suggest that the interconnection structure can be represented by or implied in tables and/or algorithms that will enable the determination of such things as the next eligible node for resource utilization, internodal lengths, reconfiguration parameters, and optimal paths. A complete representation might be underutilized in the present simulation effort, but would provide for increased sophistication in the future.

If the node is resolved into its components (i.e., the computing element and IU), it can be seen that the model can represent the interconnection aspects of the various control schemes possible for distributed networks. Since no assumption is made about the nature of the components of a node or of its communications with the rest of the network, it is possible for a particular node to represent a centralized controller dedicated solely to network management instead of a host processor in the usual sense.

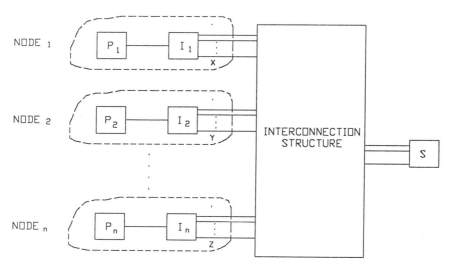

Figure 10-4 General interconnection model.

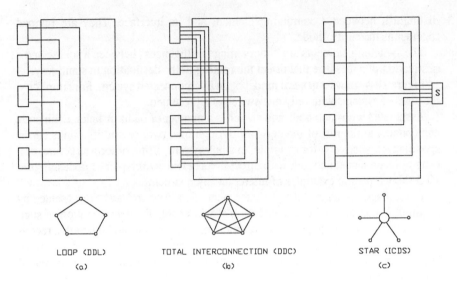

Figure 10-5 Examples of interconnection structures.

PROTOCOLS

Network activities occur in a potentially hostile environment because of such factors as nonhomogeneous components, limited bandwidth, delay, unreliable transmissions, and competition for resources. In order to provide for the orderly coordination and control of activities, formal communication conventions or protocols have been developed that encompass the electrical, mechanical, and functional characteristics of networks.

These protocols are almost always complex multilayered structures corresponding to the layered physical and functional structure of networks. Each lower layer is functionally independent and entirely transparent to all higher-level layers. However, in order to function, all higher-level layers depend on the correct operation of the lower levels.

Every time one protocol communicates by means of a protocol at a lower level, the lower-level protocol accepts all the data and control information of the higher-level protocol and then performs a number of functions upon it. In most cases, the lower-level protocol takes all the data and control information, treats it uniformly as data, and adds on its own envelope of control information. It is in the format of messages flowing through a network that the concept of a protocol hierarchy is most evident. The format of transmitted messages shows clearly the layering of functions just as a nesting of parentheses in a mathematical expression or in a programming language statement does.

Among the functions provided by protocols are circuit establishment and maintenance, resource management, message control, and error detection and correction.

Performance of these functions provided by protocols are circuit establishment and maintenance, resource management, message control, and error detection and correction.

Performance of these functions introduces delays in data transmission and requires adding headers and other house-keeping data fields to messages as well as requiring acknowledgment of correct reception or retransmission in case of errors. This reduces the useful data rate of a network. These overhead aspects of message transfer transmission are taken into account in a measure of the efficiency of the protocols. In general, "a protocol is simply the set of mutually agreed upon conventions for handling the exchange of information between computing elements." Although these elements could be circuits, modems, terminals, concentrators, hosts, processors, or people, the view taken in this section is restricted to hosts and processors embedded within other equipment.

Basically, "the crux of maintaining a viable distributed environment lies in accepting the inherent unreliability of the message mechanism and [to] design processes to cope with it." In earlier systems, protocols were designed in ad hoc fashion. Typically, these protocols were application specific and implemented as such. "All recent protocol work has been moving in the direction of a hierarchical, multi-layered structure, with the implementation details of each layer transparent to all other layers and hierarchy" (24). Although the number of layers and functions assigned to them are not generally agreed upon, Figure 10-6 shows a typical implementation. Table 10-1 describes the functions associated with each level. In practice, levels 1 through 4 are better understood and, hence, better defined, whereas levels 5 and 6 are, for the most part, still in the development if not in the research stages.

There are numerous advantages to a layered approach for protocols; they include:

1. Implementation issues are separated from functional issues.
2. It supports graceful evolution, thereby providing the flexibility to incorporate new technologies at any level without creating a traumatic impact on the layers above the affected layer.
3. It gives enhanced security resulting from well-defined and formal interfaces between layers.
4. Separation of functions should facilitate the management of the implementation.
5. There is segregation of responsibility for resource management; each layer is responsible for the management of a different class of resources.
6. It has better maintainability because of formal interfacing requirements.

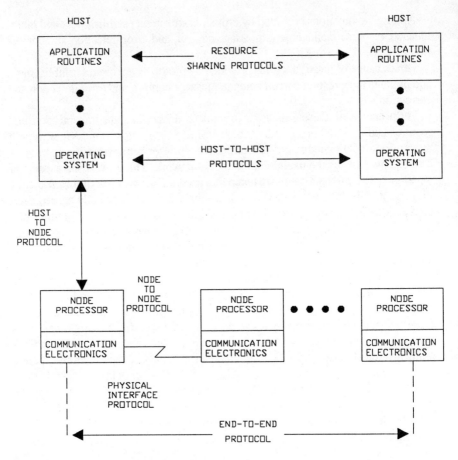

Figure 10-6 Typical model of protocol hierarchy.

There are, however, several problems associated with the design, development, and implementation of protocols; they are:

1. There is a need for a taxonomy for protocols so that:
   a. Methods can be developed to compare and analyze differing protocols.
   b. Critical features and issues of each layer can be explored.
   c. Layer definition is sharpened.
2. There is no current agreement on how to decompose protocols into the appropriate layers.
3. It is difficult to establish that a protocol is "correct" (i.e., to provide avoidance of deadlocks, etc.).
4. There is a need to establish a methodology to determine the trade-offs between providing efficiency and ensuring correctness.

Table 10-1 Description of Protocol Layers

    Layer Name   Description

1. Physical interface protocol   Mechanical and electrical characteristics/ interface
   Timing
   Definition of control
2. Node-to-node protocols   Governs data flow over physical circuit, concerned with:
   – Medium acquisition
   – Routing
   – Error control and recovery
3. Host-to-node protocol   Message assembly and disassembly into packets
   Message formatting
   Encryption Packet control
4. End-to-end protocol   Management of logical vice physical circuits or channels
   Concerned with message controls
   Maintenance of logical connections
   Routing of messages in vice packets among processors
5. Host-to-host protocols   Session control
   Provide sources to network users
6. Resource-sharing protocols   Network transparency
   Distributed data management functions
   Transparency of network resources to users

None of these problems, however, are critical in the sense that they would prevent the development of a distributed system. Rather, these issues are primarily concerned with developing better, more efficient, and reliable protocols and structures.

Although there is no universal agreement on the names and numbers of protocol layers, a widely accepted standard is that of the International Standard Organization (ISO), which is shown in Figure 10-7.

Using this organization, level 1 (physical layer) protocols include RS-232 and X.21 line control standards, Manchester II encoding, encryption, link utilization time monitoring and control, transmission rate control, and synchronization.

Level 2 (data link) provides for the reliable interchange of data between nodes connected by a physical data link. Functions include provision of data transparency (i.e., providing means to distinguish between data and control bits in a transmis-

230 Modeling and Analysis of Local Area Networks

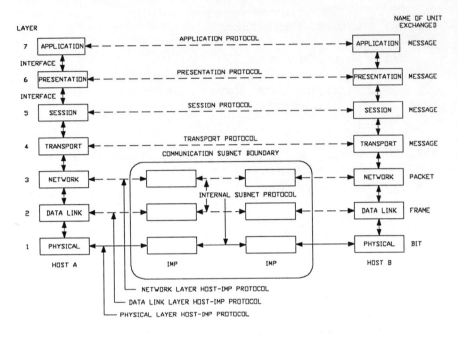

Figure 10-7 The network architecture used in this book. It is based on the ISO OSI reference model.

sion), contention monitoring and resolution, the establishment, maintenance, and termination of interactions (transaction), error detection and correction, and nodal failure recovery.

A description of the operational aspects of the general network is best presented in the context of the previously defined protocol structure since all possible network events and activities, intentional and otherwise, must be managed under this structure. The protocol structure also implies the underlying structures and functional mechanisms that support network operation.

Before any control or data communications can be conducted, the actual means of signaling and bit transmission across a physical medium must be provided. Physical links must be established in accordance with the specified network topology and line parameters.

Frequently, an encoding scheme such as Manchester II is used on this level to provide for synchronization and error detection. In the Manchester II scheme each of the original data bits is transformed into two transmission bits in such a way that it is impossible to get three consecutive identical bits in the encoded message. This implies that the message receiver can detect errors by watching for this occurrence. Also, this encoding can be selectively disabled to provide unique invalid waveforms that can be used as synchronization signals.

Given the physical layer service capability to exchange signals across the physical medium, the data link layer is implemented to provide the capability of reliably exchanging a logical sequence of messages across the physical link. The fundamental functions of the layer include the provision of data transparency, message handling, line management, and error control. Since in the original case data and control information pass along the same line during a transmission, certain techniques must be provided to distinguish between the two. This is provided by assigning control meanings to certain bit patterns that are prevented from occurring in the data stream through the use of such techniques as bit and byte stuffing, and the previously described Manchester scheme. In this way, control sequences can be used to delimit the beginning and end of asynchronously transmitted, variable-length messages. Common expressions for such sequences include BOM, EOM, and flag.

The elementary unit of data transmission is usually the word. The number of data words in a message is generally variable up to some maximum message length (MML), and a parity bit is usually appended to data and control words. Each message must include addressing information whenever the sender and receiver are not directly connected. Addresses may be physical, in which case each node has a unique address, or they may be logical, in which case each node has associated with it one or more coded sequences representing functional entities. A particular logical address may be associated with an arbitrary number of physical nodes, thus providing for single, multiple, or broadcast addressing. Address information may be contained in the data portion of a message or it may be part of the control information.

Each transaction may be considered to be either a bilateral or a unilateral process, depending upon whether or not the sending process requires a response from the destination concerning the success of the transmission. In the systems in which a choice can be made between these alternatives, the message must contain information about this choice. Response types include but are not limited to:

- * NO REPLY REQUESTED — In the case of a message being sent to a process where multiple copies exist, the issuance of an acknowledgment is undesirable because collisions would result.
- * STATUS REQUESTED — Information regarding the success or failure of the transmission is requested.
- * LOOPBACK REQUESTED — Loopback is the situation in which a destination node is also the source node.

TRANSMISSION ERROR DETECTION

In order to ensure that a transmission is occurring without error, it is necessary for the link control level to include a set of conventions between the sender and receiver for detecting and correcting errors.

There are many possible methods for error control over a transmission link. Two general types of control are: forward error control, in which sufficient redundant material is included with the information to allow the receiver to detect an error and to infer the correct message, and feedback error control, where some redundancy is needed to reveal errors, but correction is made by retransmission. The most practical and prevalent method is feedback control.

The simplest form of detection is a parity check on each transmitted character. This is often called a "vertical redundancy check," and it is used to provide protection against single bit errors within characters. A horizontal or longitudinal redundancy check (LRC) provides for a check across an entire message. This is done by computing a parity bit for each bit position of all of the characters in the message. The most powerful form of check is the cyclic redundancy code check (CRC), which is a more comprehensive algebraic process capable of detecting large numbers of errored bits.

There is a possibility that a message or response does not even arrive at its destination, irrespective of whether the information is good or bad. This can result from either a physical failure, such as the failure of the link or of the destination node, or a logical failure, such as the use of an incorrect destination name.

These possibilities can be detected by providing a time-out mechanism that will cause a message to be retransmitted if, after an agreed upon delay (the response time-out), an acknowledgment has not been received. Many systems rely solely on the positive acknowledgment and time-out convention and do not employ a negative acknowledgment.

When multiple devices are sharing a bus, there must be some method by which a particular unit requests and obtains control of the bus and is allowed to transmit data over it. The major problem in this area is the resolution of inevitable bus request conflicts through the use of arbitration and scheduling schemes so that only one unit obtains the bus at a given time. Mechanisms must also be provided for system reinitialization and adjustment in the cases of system start up, nodal addition and removal, line failures, and spurious transmissions in the system.

In all systems collisions can occur when more than one control or data transmission simultaneously occurs. This may be caused by the use of random number techniques to generate allocation sequence numbers upon a node entering the system at start up or some later time, or it may be caused by a message with multiple destinations improperly asking for acknowledgment. Collisions are usually handled by the temporary or permanent removal of involved nodes or the retransmission of legitimate messages.

A limit is often imposed on line use time to prevent a node from monopolizing the bus, either intentionally or because of nodal failure. This condition may be prevented by placing a limit on the maximum message size and/or monitoring line use to determine when a node is maintaining an active transmission state beyond that required to send the largest allowed message.

This monitoring capability is achieved through the use of a "loud-mouth" timer that is activated upon nodal allocation and that provides an interrupt signal (or causes a collision) if allowed to run out. The usual outcome is the removal of power from the transmitter circuitry, either temporarily or permanently, and the informing of the host processor, when possible, of this condition.

When talking about control, it is important to keep in mind that this is not usually associated with a specific physical device or location but is rather a functional entity distributed (replicated) throughout the network.

EVENTS

In order to precisely simulate the operational behavior of networks a more formal and quantitative analytical approach must be taken. In order to do this, the following concepts must be introduced.

All actions and activities, intentional and otherwise, that can occur in system operation can be classified as events. An event is defined as any occurrence, regardless of its duration. Events have a number of characteristics:

- An event has a beginning, an end, and a duration
- It can be simple or complex. A simple event is one that cannot or for the purposes of the simulation, need not be reduced into a simpler sequence of occurrences. Conversely, a complex event is one that consists of simpler events.
- It may be a random occurrence or of a stochastic nature, or it may be the deterministic result (effect) of an identifiable cause.
- It has a certain pattern of occurrence; e.g., periodic, aperiodic, synchronous, asynchronous, etc.
- Events belong to classes. The significance of an event class is that each member has the same effect as each other in a particular context. For example, in certain systems the corruption of a message by noise is equivalent in effect to the incorrect specification of a destination name — both will result in a response time-out and a retransmission. Thus, these two events would be of the same class.
- Events may be concurrent or disjointed (sequential). Events that coincide or overlap in time are concurrent; otherwise, they are disjointed. The concept of "effective concurrency" is introduced here. Sequential program structures are considered effectively concurrent if they can successfully represent or model events that are actually concurrent.

An example would be the action of processors requesting services. While this is an asynchronous, unpredictable event(s) concurrent with channel utilization by a particular node, these two aspects of system behavior — utilization and contention — can be effectively separated since (except in the case of interrupts) the request will not be acted upon until utilization is complete. As long as a record of the duration of utilization is available, an effective history of nodal requests can be generated just prior to contention resolution by the control module of the simulation program.

234 Modeling and Analysis of Local Area Networks

Following is a list of basic events that may be found to occur in the operation of various LCNs. All system behavior can be ultimately reduced to sequences of these simple events. Events are listed under the component in which they occur.

Processor

- Production. This is the generation of a message by a host computing element. Parameters associated with this event are production time, message size, destination(s).
- Queue inquiry. The determination by the processor of the state of an input or output queue before reading from or writing to it, respectively.
- Output message disposition. Depending upon the buffer availability strategy, a generated message may be queued normally, it may be written over exiting queued data, it may be held by the processor until it can be queued, or it may be dumped.
- Consumption (read message). This is the reading of a message in the input queue by the host computing element. Assuming a message is available in the input queue, it can be immediately consumed. Consumption with respect to a queue is similar to production.
  - Node. It is convenient to associate the following events with the entity node rather than either the computing element or the IU.
- Addition. A node is considered added to the circuit when it "informs" the network that it wishes to be integrated into the system. This may occur upon initial power-up of the node or upon failure recovery.
- Integration. This is when the node actually becomes a functional part of the network.
- Failure. This is the failure of a node as a functional member of the network.

Interface unit

- Queue inquiry. This is analogous to the processor event.
- Read message. The IU obtains message from an output queue.
- Preprocess message. This is the formatting or packing of a message for transmission. It is to be distinguished from the formatting that is done by the processor.
- Request. The IU notifies the network that it wishes to use the bus. This may or may not involve the transmission of a control signal.
- Connection. This is the actual acquisition of the channel for utilization.
- (Re)transmission. This is the moment when the first word of a message is placed on the bus or, in the event of a message train, as in a ring structure, the first word of the first message.
- Response time-out activation. In systems in which a response to a message transmission is required within a certain amount of time, a response timer is activated at some point during transmission.

- Detection (identification). This is the detection by an IU of a message addressed to it.
- Reception. This is the moment when the complete message has been received, processing on it has been completed, and it is ready to be queued. This may also be considered queue inquiry time as well as response transmission time.
- Write message. This is when a received message is placed in the input queue.
- Response reception. This is when the message source receives information from the destination regarding the transmission.
- Delete message. This is the deletion of a message from an output queue following a successful transmission.
- Relinquish. Upon completion of utilization, the IU signals that reallocation is to occur.

Using these concepts, overall system activity or flow can be represented by the following sequence of complex events:

NODAL ACTIVITY —>CONTROL/ARBITRATION —>UTILIZATION

This simple structure is possible because the concept of effective concurrency is valid in the case of global-bus systems.

Nodal Activity

Nodal activity simulates the behavior of all nodes and interface units during the utilization period of a particular node. During this period nodal activity includes:

- The production and consumption of messages
- Queue activity
- IU background processing
- Addition and removal of nodes from the network
- Etc.

Control

Control may be a number of possible sequences depending on circumstances in which control is activated.

Utilization

The utilization event encompasses all activity associated with a node's utilization of the bus for the transmission of a message. Utilization begins with being "connected" to the bus and it terminates either gracefully, in the case of a successful transaction followed by a control output, or unintentionally as the result of intervention by control because of a protocol violation (message too long, no reallocation signal transmitted, etc.).

THE MALAN MODEL STRUCTURE

Based on the previous discussion of LAN structure and events, we can typify a LAN as consisting of three major hardware classes: host machines, interface units, and communications links. Additionally, these hardware classes possess varying levels of software and functionality.

The link level is concerned with the management and performance of bit-level physical transfers. This includes timing and control as well as mechanical and electrical interface. The interface units provide the main services to bring the simple communications media and protocols up to a true "network." This component and its services must provide for node-to-node, host-to-node, and end-to-end protocols. This includes error detection and correction, media acquisition and control, routing, flow control, message formatting, network transparency, maintenance of connections, and of other services, as discussed in Chapter 2. The host class of device provides the LAN with end-user sites that require remote services from other hosts. The services provided at this level are host-to-node interface, host-to- host protocols, and resource-sharing protocols. Implemented at this level of a LAN would be user-visible services such as a distributed operating system, a distributed database management system, mail services, and many others. A model of this structure implies a minimum of a component for each of these items. Therefore, MALAN must have components of sufficient generality and flexibility to model these components and provide for analysis. Figure 10-8 depicts these basic components and the necessary simulation components to measure the performance and opera-

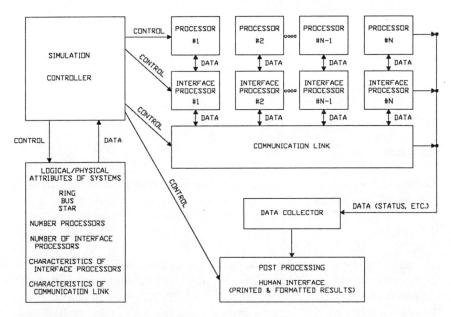

Figure 10-8   MALAN initial structure.

tions of a simulation. Using this structure, it can be seen that MALAN consists of a modular structure with components that can be turned to the modeling of specific LAN nuances. The physical and logical characteristics peculiar to each system design are contained in independent software routines and/or data tables. The high-level design of the LAN model shows the need for the following functions:

- A simulation controller that will be responsible for the coordination and timely operation of the remaining software modules. This function will initialize the system architecture and distributed computing techniques in accordance with user input data, schedule events, calculate system state, maintain the common timetables, and initiate the processing of the other software routines as dictated by the particular logical and physical configuration.
- A system processor routine that will be capable of simulating the time-dependent activity (data in, data out, processing time) of each proposed computing mode.
- An interface processing routine that will be capable of simulating the activity (time delays, message handling, priority determination, addressing technique, resource allocation, etc.) of each proposed front-end processor as required by the particular logical and physical characteristics.
- A communications link routine that will be capable of simulating the timing delays and the data and control transfer characteristics of the proposed transmission medium.
- A data collection routine that will be responsible for collecting, formatting, and collating the requisite system evaluation parameters.

Data items collected will include, but will not be limited to, a minimum, maximum, and average of the following:

- Time to transmit message from A to B
- Message wait time
- Number of messages in the queue or system
- Message size
- Bus utilization
- Interface unit timing (as previously presented)

It will also include a post-processing routine that will be responsible for presenting the data in human readable forms, (graphs, plots, tables, etc.).

MALAN SIMULATOR OVERVIEW

MALAN was developed to provide a flexible research, development, and analysis tool for local area network architectures. The tool has been used to aid in the selection, development, and evaluation of local area network architectures that support large distributed real-time command control and communications (C3) environments.

MALAN was designed with the intention of comparing a wide range of possible distributed C3 configurations. This capability was achieved by providing:

1. A modular structure that allows the model to be adapted to suit a variety of system specifications
2. A standard driving routine that mimics the communications within a C3 system,
3. A standard routine that analyzes the distributed system on the basis of detailed evaluation criterion

Such a design allows for the implementation, testing, and evaluation of new strategies for improving system performance with little effort.

NEXT EVENT SIMULATION

The modeling technique used in the MALAN software simulation is called "next event simulation." Next event simulation views the world as a sequence of events rather than a continuum. If a department store checkout line is simulated in next event simulation, the process of checking out would be viewed as the following sequence: (1) a customer enters the checkout line, (2) the customer starts checkout, (3) the customer completes checkout. Between these events the customer is performing other activities, but these are unimportant if we are simply interested in the length of the waiting time.

The view of time taken by next event simulation is important in understanding the design and implementation of the MALAN simulation. Time in next event simulation is viewed as a means of sequencing events and calculating time-related statistics. Events 1 hour or 1 second in the future are treated identically. Simulation is achieved by creating a file that contains future events along with the time of their occurrence. A simple loop program scans this file and selects the event with the lowest time. At that time, an internal memory location, which contains the simulated time, is updated to the occurrence time of the event. After the event occurs, mathematical calculations or logical operations can be performed to schedule other dependent events. In the checkout line example, this means that when checkout begins, the end of the checkout event is scheduled. In this way, the simulation proceeds from event to event and time constantly progresses.

The random or stochastic nature of scheduled events gives the simulation the characteristics of a real system. In the checkout line example, the time taken to serve a customer is not a constant; it may be 1 minute or 10. The service time is also randomly distributed. That is to say, the service time of previous customers does not have any effect on future customers. This means that the present customer may be followed by a customer whose service time is selected from a range of possible service times. The service times usually fall into a pattern; that is, it may be highly likely that a service time is 5 minutes but relatively unlikely to be 20 minutes. The likelihood of certain service times can be described by theoretical patterns called "distributions." These distributions can be used to generate service times or arrival patterns that resemble those occurring in real systems.

In a very simplistic view, a distributed C3 system resembles the previous example. Messages are generated by the components of the C3 system according to some distribution. They line up waiting for the service. Messages are serviced by the communications network and arrive at the destination processor.

GENERAL MODEL IMPLEMENTATION

The general model implementation is shown in the data flow diagram in Figure 10-9. which shows the flow of information in the system. In Figure 10-9 we see a user control file, random variates, and optional operating system data supplying information to the simulation and the output of information in the form of a final report. The user control data file supplies all the information that is required to configure the model into a unique distributed processing system, as well as the test data that will drive a simulation run. The random variates or random numbers supply the stochastic or random nature of the events that occur in a distributed processing system. Operational systems data are an optional input that can drive the simulation with data generated by a real C3 system. The final report consists of quantitative information that summarizes the functioning of the system. This report will supply sufficient information to meet the goals of the MALAN software simulation.

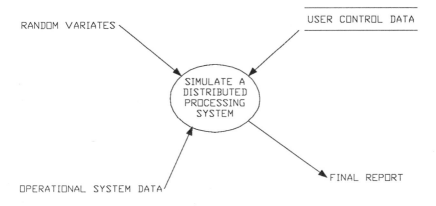

Figure 10-9 General model high-level data flow diagram.

MODEL IMPLEMENTATION

The simulation model has been made to resemble the generalized conceptual model of the distributed processing system shown in Figure 10-10. The processing element in the Figure 10-10 initiates and receives information. The information to be communicated is shown formatted into discrete packets or messages. Messages M1 and M2 are waiting in a line or queue for service from IU, which acts as the intermediary between the processing elements and the vehicle of communication, the communications network. The IU removes messages from the queue, transforms them into a form suitable for transmission, acquires the use of the communications network, and transmits the message. In reality, distributed processing systems are composed of a number of these processing and IU units connected to a network. Thus, another IU in the system is prepared to receive the message. The receiving IU draws the message off the communications network, formats it for the processing element, and completes the process by making the message available to the processing element. The received message is shown as message MO.

The MALAN software simulation of the generalized conceptual model is shown in Figure 10-11. This is a more detailed version of Figure 10-9 with the user control

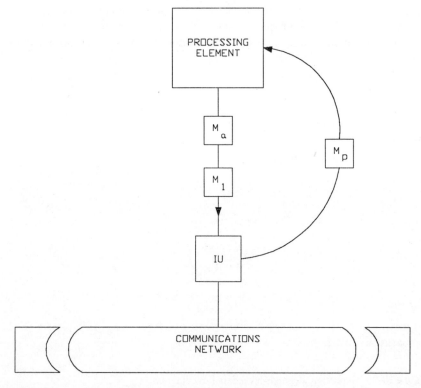

Figure 10-10   Generalized model of a distributed processing node.

data file excluded. This was done for simplicity, but it should be noted that the user control data file has input into each one of the processors shown in Figure 10-11.

Each of the bubbles in Figure 10-11 describes a major functional component of any distributed processing system. In general terms, each distributed processing system requires: (1) processing elements as a source of arrivals and the ultimate consumers of communication; (2) a queuing mechanism that stores messages waiting for transmission; 3) a method for arbitrating or allocating the limited resources of the communications system; (4) a communication network that physically transmits the message; and (5) a system that routes messages through the system. For simulation purposes, a 60-component system is required. The analysis module performs the analysis required to evaluate the performance of the system and issue a final report (this acts like a monitoring device would on a real system).

Each of the process bubbles shown in Figure 10-11 has a subsequent section that contains more detailed descriptions of its function. In this introductory section, each will be described in general terms along with its interaction with other processes.

The arrival module replaces the processing element in the generalized conceptual model. The function of this process is to take random variates or operational systems data and produce message arrivals. These message arrivals are essentially small buffers of computer memory that can be passed around the system, mimicking the movement of real messages. These simulated messages contain information that characterizes the message as to its origin, size, etc. During the passage of the message through the system, historical information will be attached to the message. It is this information that will be used to characterize the passage of the message from source to sink processor.

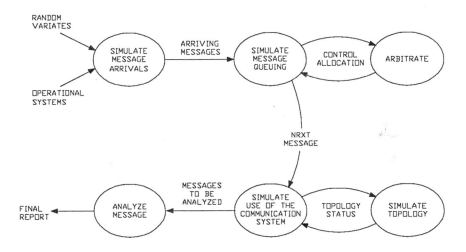

Figure 10-11 Software implementation of generalized data processing system model.

The message queuing process accepts the message arrivals and places them in a waiting line or queue corresponding to the processor of origin. Allowances are made for the fact that queue memory may be limited and overflow is possible. It is not obvious from the general conceptual model, but the communication network is generally a limited resource. In order to use the communication network, this resource must be allocated. This function is performed by the arbitrator module.

The general conceptual model depicts the communication network as essentially a black box. Actually, it may consist of one or many interconnections between IUs. To guide messages around this network, a position keeper and "road map" are necessary. This routing and position-keeping is the function of the topology module.

The use process is responsible for simulating all the delays and other procedures that are part of the physical communication process. This module takes the message selected by the arbitrator and simulates its transmission. The module also changes the historical information contained in the message for future reference.

It is the function of the analysis process to take the historical data from each of the messages moving through the simulation and accumulate statistical data. At the end of a simulation run, these accumulated data are formatted into a final report that analyzes the outcome of the simulation run.

MALAN INTERACTIVE SIMULATION INTERFACE

The user control data file supplies all the information that is required to configure the MALAN software simulation model to represent a unique distributed processing system. The user control data file will also supply test data that is used during a simulation run. For the initial implementation of MALAN, the user data file will be laboriously assembled by hand for each simulation run. Much of the work is forming tabular data from higher-level descriptive information. For the initial phases of the model testing and experimentation, the hand assembly of data is adequate.

To reduce the labor required to use the system for experimentation and testing of distributed processing systems, a more friendly interactive user interface will be built. The interactive simulation interface, as it is called, will prompt or query the experimenter for information that completely describes the details of the experiment. The query will be in the form of menus, listing alternate selections, presented on a CRT screen. User response to the menu queries will be accepted using the CRT keyboard; these responses are called user data. The interactive simulation interface will also test the "user data" for validity of the type and range. Validity testing will help avoid errors in the input parameters and increase confidence in the resulting data.

The basic design of the interactive simulation interface is shown in Figure 10-12. This figure shows the user receiving query data and supplying a response. The response is processed by the 1.0 configure simulation run to yield test data that are

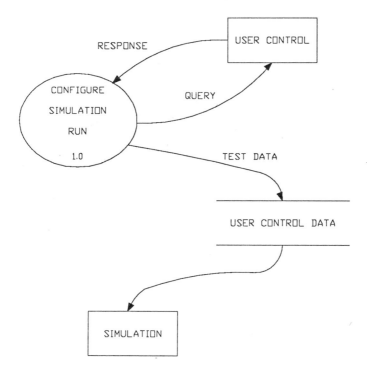

Figure 10-12 High-level interactive simulation interface.

stored in the user control data file. This file is used by the simulation to create the operating environment for a simulation run.

In this particular implementation of MALAN, the user control data consists of 80 ASCII encoded control records. The type of data that is contained in the file can be divided into (1) control data that are used by the GASP simulation language to configure GASP functions and (2) control cards that are unique to MALAN.

Figure 10-13 further breaks down the interactive user interface into seven separate configuration processes. These configurations are closely aligned with information required by the major modules of the simulation. The data generated by the separate configuration processes 1.1 through 1.7 are compiled into a single-user control data file by process 1.8 compile test data.

MALAN MODEL IMPLEMENTATION

MALAN is a general purpose simulation package usable for modeling a wide range of local area network architectures. Its basic structure is shown in Figure 10-14 and consists of five major components:

244 Modeling and Analysis of Local Area Networks

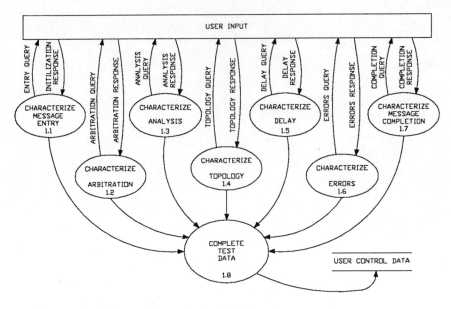

Figure 10-13 Low-level interactive simulation interface.

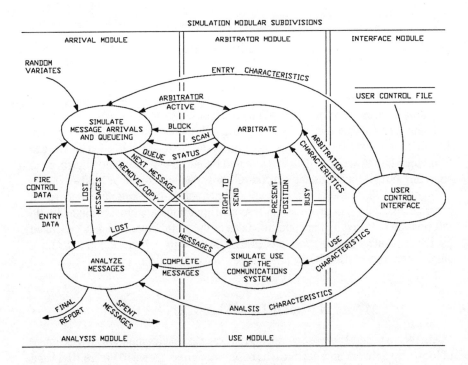

Figure 10-14 MALAN simulation modular subdivisions.

1. Arrival module. The arrival module is concerned with generating the messages to be communicated and places them in an interface unit queue. It must also handle the queue overflow problem and the possibility of a processor being unavailable.
2. Arbitrator module The arbitrator module is concerned with the determination of which interface unit will communicate over the link next, based on the policy of the communications link control in place.
3. Use module. The Use module is concerned with the modeling of the passage of messages from source to sink nodes over the communication link.
4. Analysis module. The analysis module performs statistical analysis on the messages within the system.
5. Interface module. The interface module handles overall control of the characteristics of the simulation.

Table 10-2 lists the data links from Figure 10-14 and provides a high-level meaning for them.

One of the critical modules of MALAN in terms of general characteristics modeling is the use module. A detailed view of the data flow and components for the use module is shown in Figure 10-15. Its major components are:

1. ADD NEXT MESSAGE TO THE FILE. This process takes the next message from the processor that has the right to send and stores it in a file until it arrives at its destination.
2. SIMULATE TOPOLOGY. This routes the message through the simulation topology.
3. SIMULATE TRANSACTION DELAY. It simulates the passage of a message over a physical transmission line (LINK) including messages retransmitted.
4. SIMULATE MESSAGE ERRORS. This determines the number of retransmits and lost messages from information on the physical transmission line.
5. UNFILE MESSAGE. This process removes messages from the message file upon completion of message transmission.
6. SIMULATE STATUS CHANGE. This simulates the loss of a link or node.

The data flow meanings are shown in Table 10-2.

ARRIVAL MODULE

In a general sense, the mandate of the arrival module is to simulate the generation and queuing of messages within a C3 system. The function of the arrival module can be further broken down into two subfunctions: simulate message arrival and simulate message queuing.

Table 10-2 MALAN High-End Data Flow

Data link	High-level meaning
Add Characteristics	Specifies the characteristics of a message in transit, such as headers, etc.
Analysis Characteristics	Specifies the types of graphics and tables to be generated.
Arbitration Characteristics	Specifies the type of arbitration to be performed.
Arbitrator Active	The state of the arbitrator is active; i.e., in the process of servicing the queue; when inactive, it is waiting for an arrival to start servicing the queue.
Arbitrator	Data statistical information related to arbitrator activity.
Bit Loss	Indicates that a message failed to complete due to a transmission error.
Block	Specifies the processors that are blocked and unable to transmit, and those that are unblocked and able to transmit.
Busy	Signals that the system is in a process of transferring a message.
Completed Messages	Successfully completed messages.
Completion Characteristics	Specified the way messages enter a receiving IU.
Delay Characteristics	The characteristics of a system delay.
Destination	The destination of the message taken from the message attributes.
Entry Characteristics	Arrival characteristics plus request characteristics. The characteristics of non-fire control load plus a specification of the action to be taken when a message is placed in the queue.
Entry Data	The total number of messages entering the system plus the number of messages in the queues plus statistics on blocking.
Error Characteristics	Specifies the errors encountered over the physical transmission lines.
File Contents	Keeps an account of the message/ messages contained in the message file.
Final Report	Graphic and tabular data that analyzes the simulation run.
System Control Data	Data that regulates the generation of messages to more closely resemble those generated by a real system.
Lost Messages	Messages that are lost by the system.
Lowest Message	Messages that are lost by the system.

Table 10-2 MALAN High-End Data Flow (cont.)

Data link	High-level meaning
Message File	Contains a message in transit.
Next Message	The processor number of the processor that will transmit next.
Present	Position specifies the processor that will gain control next.
Queue Status	States: 1. IU contains a message. 2. IU empty, no message. 3. IU blocked, not able to send.
Random Variates	Random numbers between which are used to generate stochastic nature of the model.
Receiving Processor	The process or to receive a message.
Remove/Copy	A request for the next message to be removed or copied from a specified queue.
Retransmits	The number of time a message is retransmitted due to error.
Right to Send	The process or number of the processor that can send a message.
Scan	Contains the state of an IU.
Spent Message	Message data that is no longer relevant to the system.
Status Change	Indicates the loss of a link or a node during the simulation.
Status Loss	The loss of a message due to a status change.
Success	Indicates a successfully completed message.
Topology Characteristics	Defines the interconnection structure of the communication system.
Topology Status	The status of the physical wire over which the transmission is carried and the processor to which a message is transmitted.
Topology Tables	The data that contains the topology characteristics of the system under test.
Total Transmission Size	The total amount of information contained in the message file.
Use Characteristics	Add characteristics plus topology characteristics plus delay characteristics plus error characteristics, plus status changes plus completion characteristics. The way messages are transmitted by the system.
User Control File	The file that specifies the type of communication system to be simulated, the desired inputs and outputs.

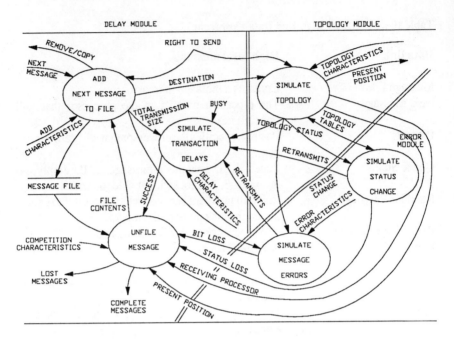

Figure 10-15 Use module.

Simulating message arrivals requires a source of message interarrival times. An interarrival time is defined as the period between two adjacent arrivals: tn and tn + 1, where tn is the time of the Nth occurrence and tn + 1 is the time of the n + 1st occurrence. The generation of interarrival times will be discussed in subsequent paragraphs; for the time being, interarrival time can be defined as simply the time between message arrivals.

In order to understand how interarrival times are used to simulate message arrivals, a description of how the simulation starts and how messages flow through the system follows. To start the simulation, an arrival event is scheduled for each interface unit. Scheduling consists of adding the present simulated time to the interarrival time to yield the event time or the time of actual message entry into the system (see Figure 10-16). In order to simulate the occurrence of the arrival event, the event time is entered into a file of all future events. During the simulation, this future event file is scanned and events such as arrivals are made to occur in their correct temporal sequence.

An arrival event occurs when the event time of a particular arrival in the file of future events matches the present simulated time. Simulated time is maintained by a software clock that is incremented to ensure the proper temporal sequence of events.

The occurrence of an arrival event is a signal to the system that a message generated by the simulated C3 system is ready to enter the system and attempt transmission. At the time of a message arrival, another message is scheduled for the receiving IU by again invoking the interarrival time generator. This practice of

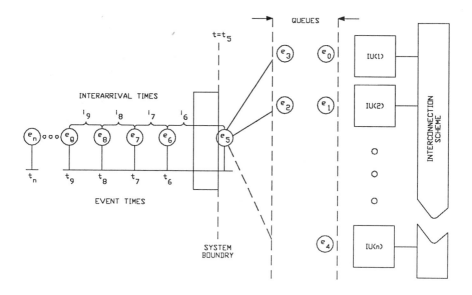

Figure 10-16 Arrival module.

triggering subsequent arrivals from latest arrivals ensures a constant chain of messages entering the system.

There are several ways to generate interarrival times for simulations of this type. The first method is to measure interarrival times of messages generated by an actual running C3 system. Properly generated data of this type can be used to answer the question "how would the present system function on a distributed network of type X?" As the system is running, interarrival times are stored on some input media such as magnetic tape. As the simulation progresses, the values are read from tape and arrivals would be scheduled appropriately. Another perhaps more flexible method is to generate the interarrival times by a theoretical distribution. Theoretical distributions are mathematical relationships that are designed to closely approximate the distribution generated by the real system. Additional flexibility is afforded by an ability to change the theoretical distribution slightly in order to test some worst-case situations or some distribution that might exist in a future system.

The simple arrival of a message is not sufficient to simulate the generation of a real system message. Real messages have several additional features: (1) they have a specific size and (2) they have a specific destinations. The system needs this information to determine the path and delays a message will encounter when passing through it. For example, a large message will require longer transmission times and may be more susceptible to error. Also, messages bound for one IU may be required to take more intermediate transmission time than another. To facilitate these characteristics, a message shall consist of a buffer of information that represents a specific message type. That is to say, computer memory has been allocated for each message and information is placed in this memory in specific places to allow easy data retrieval. Within the message type there are data that identify the origin, destination, and size of a message. The message type also contains information regarding the passage of the message through the system. The analysis module is concerned with these attributes since they are used to yield the final statistical results.

At this point, a discussion of some of the parameters that form the message type is appropriate (see Table 10-3). Each of the items in the table is an attribute. Attribute 1 is the event time; it contains the time at which the message will arrive. Attribute 2 is the event type, that distinguishes this event as an arrival, since there are other events that occur in the system. These other events will be discussed in subsequent sections. Attribute 3 is the source IU number or the designated IU that generated the message. Attribute 4 is the destination IU number and describes the ultimate destination of the message. Attribute 7 is the message size expressed in words. Attributes 17 and 18 are used when a message is too large to be sent in one packet. In this case, the message must be divided into a number of smaller

Table 10-3 Message Data Types

ATTRIBUTE NAME	DATA NAME	DATA TYPE	ABBREV- VIATION
1	EVENT TIME	REAL	ETIM.
2	EVENT TYPE	REAL	ETYP.
3	SOURCE INTERFACE UNIT (IU) NUMBER	INTEGER	SP.
4	DESTINATION INTERFACE UNIT (IU) NUMBER	INTEGER	DP.
5	PRESENT INTERFACE UNIT (IU) NUMBER	INTEGER	PP.
6	GENERATION TIME	REAL	GT.
7	MESSAGE SIZE (WORDS)	INTEGER	MS.
8	MESSAGE OVERHEAD LENGTH (BITS)	INTEGER	MO.
9	MESSAGE WAIT TIME ΔT 1 (IN PROCESSOR QUEUE)	REAL	WT 1
10	MESSAGE WAIT TIME ΔT 2 (TRANSIT FROM QUEUE TO IU)	REAL	WT 2
11	MESSAGE WAIT TIME ΔT 3 (WITHIN IU)	REAL	WT 3
12	MESSAGE TRANSMIT TIME	REAL	TT.
13	MESSAGE TRANSFER TIME	REAL	XFER.
14	NUMBER OF STOPS	INTEGER	NS.
15	NUMBER OF RETRANSMITS	INTEGER	RT.
16	MESSAGES LOST	INTEGER	ML.
17	SEQUENCE NUMBER OF MULTI-PACKETED MESSAGES	INTEGER	NMES.
18	NUMBER OF PARTS TO A PACKETIZED MESSAGE	INTEGER	PARTS.
19	MESSAGE TIME TO COMPLETE	INTEGER	MTTC.
20	MESSAGE PRIORITY	INTEGER	MP.
21	MESSAGE IDENTIFICATION (ID) NUMBER	INTEGER	MI.

(WT 1, WT 2, WT 3 grouped as WT)

messages, each with the same source, destination, and generation time. Attribute 17 is used to identify the sequence number of any multipacketed message. Attribute 18 is used to identify the total number of packets in the entire message. With attributes 17 and 18, it can be determined when the complete message is received.

Like interarrival times, message size and message destination must be generated by the system. Here again, the data could be generated by measurements within the real system or by using theoretical distributions. Thus, part of the arrival module is devoted to drawing from distributions of message size and message destination and initializing the appropriate attributes. The only remaining function performed by this submode is to divide messages that exceed the maximum message size into a sequence of smaller messages. The total number of messages generated is placed in attribute 18 of each message. Each message receives a sequence number, attribute 17. At this point, all attributes that define a simulated message have been defined (see Figure 10-17).

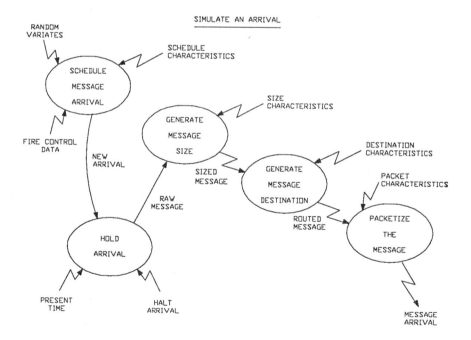

Figure 10-17   Attribute definitions.

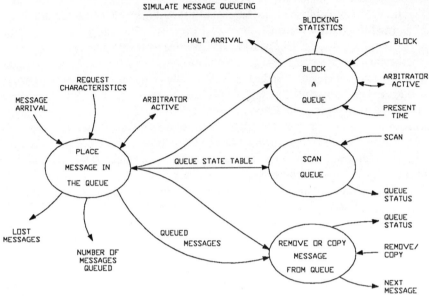

Figure 10-18 Simulate message queuing.

Once messages enter the system, they must be held in a waiting line or queue until the IU can transmit them (Figure 10-18). To facilitate this, a FIFO queue is formed to contain the waiting messages. In a real system, this queue would consume some real memory that normally would be limited. In the simulated system, this limit must be taken into consideration. When queue memory is exceeded, appropriate action should be taken.

Appropriate action consists of: (1) waiting until sufficient memory is free to accommodate the message, (2) throwing away the message, and (3) overwriting an older message. Any of the above can be selected in this submode.

Simulate an arrival. Also, part of this module contains the logic that allows the rest of the system to access messages in the queue. In the real system, messages are read from the IU queue and transmitted through the system. A similar function is provided by program logic within the simulation. The major interfacing functions are copy, remove, scan, retrieve, and restore.

Copy allows the first message in the queue to be made available to other components of the system, presumably to transmit it through the system. Copy frees the memory occupied by the message, but retains a copy of the message in an interim file. This copy simulates a holding area that exists in many IUs. It will remain there until removed, presumably when successful transmission is completed. When a copy is performed while a message is in this interim file, the interim message will be made available. A status indicator is also available that supplies information such as the status of the interim file, the presence or absence of a message, or the contention block condition; these will be discussed in subsequent paragraphs.

Remove serves the same function as copy, but removes the message from the queue or interim file. Status information is also supplied.

Scan simply supplies status information. No messages are removed or made available. Scan is destined to allow the queues to be tested for available messages.

Retrieve and Restore are provided to allow the interim message to be manipulated and restored to the interim file. This program logic might be used to alter statistics related to attempted transmission through the system.

There is one other feature to the queuing submode; that is the contention block feature. A contention situation may occur in some communication systems that allows several IUs to contend for the communications network during some specified period. If two IUs attempt to grab the communication network simultaneously, a so-called collision occurs. Since both IUs cannot use the system simultaneously, some distinction must be made. The solution that is imposed on the system is to turn off each of the colliding IUs for an interval of time determined by a random number. At the end of this period, an IU would again contend for the communication network. During the contention period, the queuing submodule marks the IU unavailable, thus preventing any messages from leaving the queue, essentially turning it off. The blocks are removed when the contention period is lapsed.

ARBITRATOR MODULE

The general purpose of the arbitrator module is to simulate the control and allocation of the simulated systems communications resources; i.e., the data bus. The functions of the arbitrator module can be further broken down into subfunctions such as protocol, control, and allocation since all will occur within the module in some manner. Simulating control actions in a distributed system requires scheduling control arbitration, control mechanisms, and transmissions to occur. In this simulation, the following actions must occur or be present in order to perform the arbitrator's function. To start simulation for arbitration of resources, an arbitrator event must be scheduled to occur either now or at some future time, dependent on simulated system states such as message arrivals or network timeouts, etc. Once the initial arbitration event has commenced, future arbitrations will be scheduled for a future time, dependent upon the mechanisms being simulated; i.e., polling, daisy chain, contention, centralized, distributed, etc. Events are then filed away in the event file to be called upon for action in their correct temporal sequence. When the simulation controller recalls the arbitrator event to occur, the following sequence of subevents occurs to properly simulate the functioning of this module.

Control of the simulation is passed over to the arbitrator module. It then determines the mode of control; i.e., centralized or distributed, by checking the user file space. Next, the method of resource allocation and control passage is sought through a second interrogation of the user file. Control types include polling, daisy chain, and request/grant. From this point, the arbitrator will request the location and status of the last controlling user element in the network from the topology module. Once the arbitrator has a reference point, it enters one of the control mode routines and performs the actions to simulate that mode's requisite scheme. If the control mode type is polling, the arbitrator must run through the possible polling schemes in order to determine which unit will receive control next. The possible methods include round robin, in which control is passed from one logical unit to the next in a circular fashion (logical implies that units are not necessarily physically located next to each other); prioritized, in which control is always started at the controlling node and physically branches out to the farthest unit one at a time, thus giving priority to the units closest to the controlling node; update counters, which is a method by which control is determined through the updating of internal counters in each unit and when matched to a predefined value, gives control to that unit; or poll codes, which is a method through which the unit that last had control will calculate the code for the next user unit through a self-contained algorithm. A unit in the system will recognize this identifying code and take control of the communication subnetwork. If the control type is daisy chaining, the arbitrator module will scan the possible schemes available, choose the proper user-defined subtype, and compute the next controlling unit in the following manner. If a unit is requesting service, the control will pass from one unit to the next until the requester gets control. This always starts at the controlling node. If token passing is used, control will pass in a round-robin fashion from one unit to its physical neighbor, thereby allocating resources in a circular fashion.

If control is based on a request/grant or contention scheme, the arbitrator has many more functions to perform. It interrogates the active node file to determine if any unit(s) are requesting service. It next determines if the communication resource is free and responds accordingly. The arbitrator must next determine centralized or distributed submodes and perform the allocation based on this method. Centralized request/grant systems allocate distribution resources much as today's large time-sharing systems would in allocating CPU time. Distributed request/grant or contention has multiple types of control mechanisms and the arbitrator must determine the subtype; i.e., pure contention, p-persistent, persistent, nonpersistent, and multiple servers and then act accordingly to choose the next controller. Once this controlling unit has been determined, each method alluded to above will calculate its specific time to complete the arbitration cycle and check to see if there is a message to send by the new controller. If required, it will reschedule the arbitrator and schedule the use module to simulate message transmission and error conditions as required by the particular state of the system and sending unit. Once the arbitrator has completed its required processing, it will once again return control to the simulation controller, i.e., GASP IV main program, and wait for the next scheduled call.

USE MODULE

The use module is a group of subroutines and functions that function collectively as that component of the simulation process that models the communications process in local computer networks. As distinguished from the two other major system modeling components, the arrival module and the arbitrator module, which provides for the generation of messages and the resolution of nodal competition, the use module is responsible for simulating all activity related to the actual transmission, propagation, and reception of messages in accordance with system-dependent communication protocols. This includes:

- The activity of a source node during its allocation period.
- The activity of nodes receiving and processing messages or responses, which may occur at any time.
- Background processing performed by nodes while they are waiting for control.
- The passage of messages over the interconnection channels and the possible degradation of these messages.
- Intentional and unintentional changes in the operational status of the nodes.

Proper statistics are maintained during all phases of the utilization process. When the simulation of the passage of a particular message through a system has been completed, the message is "unfiled" via the simulation's analysis module.

Viewed as a GASP entity, the use module integrates and coordinates all structures and activities that constitute the simulated communications process in the framework of GASP events (see Figure 10-19), which are independent with respect to each other as well as to the events of all other modules of the simulation. Events are independent if they can affect only future events. No two events whose duration overlaps can have any effect on each other. For example, message arrivals in the system using selection channel access techniques can be scheduled to occur at any time. Arrivals that happen to have been scheduled to occur during a utilization event are queued by the GASP event-filling system and are run upon the completion of the particular use module component event being executed. In random access (contention) systems, message arrivals do affect utilization and, therefore, the overlapping of arrival and use events is precluded in the modeling of such systems. (Note: In all cases, events are never executed simultaneously. It is the time frame of events that may overlap and only then if the events are independent).

THE ORGANIZATION OF THE USE MODULE

As with the simulation as a whole, the use module is general enough to efficiently support the faithful modeling of a wide variety of systems while simultaneously

256 Modeling and Analysis of Local Area Networks

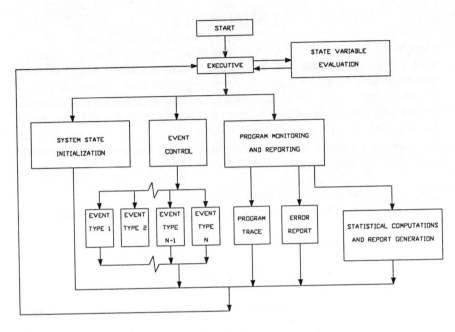

Figure 10-19 Basic modes of GASP IV control.

being capable of simulating the particular system chosen for simulation to the requisite degree of accuracy.

The Use Module represents the communications process in terms of independent message transactions originating at the network nodes. A transaction is defined, for the purposes of the simulation, as the protocol-governed activity of the source and all receivers with respect to a particular message. Each transaction is subject to various delays and errors as it is being carried out.

Each transaction is modeled using an arbitrator-initiated sequence composed of three basic events that represent the three distinct phases of any message transfer; namely, (1) message preprocessing and transmission; (2) reception and response; (3) response processing and retransmission. The exact sequence that is followed in any given case is dependent on system architecture, protocols, and the circumstances existing during the transaction. For example, the absence of a message for transmission will result in nothing more than a slight source delay and a return of control to the arbitrator, with no scheduling of any further events. A cyclic repetition of reception and retransmission may occur, if a message repeatedly fails to arrive without error at its destination. Delays and errors are introduced into the transaction when and where appropriate.

As a software structure, the Use Module consists of three main event subroutines — (1) USEINIT, (2) RECEPT, and (3) RESPRO — that correspond to the aforementioned major transaction phases:

USE INIT, scheduled by the arbitrator, is the initial event in each utilization sequence. It consists of three main parts, the imminent collision processor, the preprocessor, and the transmitter. They are:

- IMCLPR — the imminent collision processor, provides for the modeling of the activity of multiple nodes that are given control during the same allocation period.
- PREPRO — the preprocessor models all activity, excluding that related to message reception, that has taken place at the selected node from the end of its last allocation period until the transmission of a message or the relinquishment of control.

XMIT is the initial transmission of the message. The primary function of this routine is to schedule the reception event(s) at the destination node(s).

RECEPT, the reception event, models the reception and processing of a message by the destination interface unit. When mandated by conditions and protocol, a simulated response is formulated and a response event is scheduled.

RESPRO — the response even models the processing of the response by the source. This includes the possible retransmission of an unreceived or unsuccessfully received message. A collection of delay and error functions provides a pool from which the delays and errors, appropriate to the modeling of a particular system, are chosen. The scheduling relationship of the use model to other simulation events is illustrated in Figure 10-20. As can be seen, the use module is scheduled only by the arbitrator, and schedules the arbitrator, in turn. In some cases; e.g., in those systems in which arbitration is occasioned only by a message arrival, the use module schedules nothing at all, simply passing control via GASP to the next scheduled event, whatever that may be.

The use module required data from and supplies data to other system modules. These data may be in the form of global variables or passed parameters. Data exchanged may be uni- or bilateral, depending on the module involved. The arrival module furnishes information concerning message availability, size, destination(s), and other message characteristics. The arrival module receives no information from the use module. The topology module provides information about internodal distances, propagation speeds, data rates, and node status. The system status change component of the use module randomly causes changes in certain Topology data.

When the use module has determined the final resolution or outcome of a particular transaction; i.e., the disposition of a message, it notifies the analysis module of the resolution category (success, lost message, etc.) and provides current message attributes that are used for the updating of system statistics.

Internally, the USEINIT event schedules the RECEPT event unless there is no message to be transmitted, a collision is predicated, or a reception node is found to be nonoperational. If any of these fault conditions exists, the use module will immediately relinquish control. The RECEPT event schedules either RESPRO, the response processing event, or no event at all when no response, either implicit or

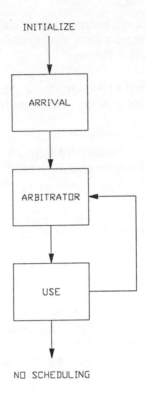

Figure 10-20 External event links.

explicit, is expressed. RESPRO either reschedules RECEPT or, if protocol allows no further retransmissions, relinquishes control.

Topology Module Description

The topology module consists mainly of a group of subroutines that facilitate the retrieval and modification of the data that physically describes the local computer network. These routines reside within the body of the simulation program with the exception of the initialization program, which is a separate entity (Figure 10-21). The following discussion will briefly describe the operation of each routine followed by a cursory view of the module as a whole. Presently, there are eight call types which may be made to the Topology Module as listed in Table 10-4. A brief description of each, along with the associated support routines, follows.

CDST

The call to the topology module for control distance is made by the arbitrator to obtain the actual distance control and must pass in a control transfer operation. The

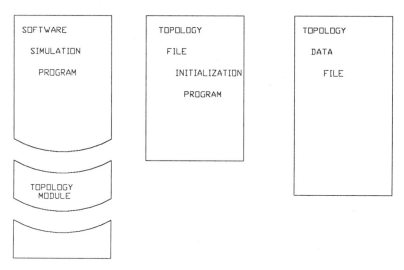

Figure 10-21 Simulation file area contents.

routine CDST obtains its input information from the global variables representing the present controlling processor (PP) and the next controlling processor (NP). Using these two inputs as source and destination, the variables SRC and DST become initialized. A procedure called "Route" is then invoked by the CDST routine. The Route routine uses the node status array (NSTA) along with the link table (LINK) to construct the current links table (CLINKS).

This operation involves the fill-in of an array with all active node links, thus creating an up-to-date representation of the system. Now with the actual system structure known, the shortest route algorithm (SHRTE) is invoked, and it calculates the various possible routes that control may follow from point A to point B. This is done by manipulating the entries contained in the CLINKS array and replacing

Table 10-4 Topology Module Calls

NAME	FUNCTION
CDST	Continue distance request
COLIDE	Collision query request
CNVRT	Convert logical to physical address
RMVND	Remove node from system
ADDND	Add node to system
DIST	Distance request
DVAL	Destination validity request
BUSTO	Bus time-out routine

them with cumulative distance values. Upon completion of the SHRTE task, the Total routine is called to read the completed system map. Total returns the control distance in the variable internodal distance (IND). Program control is then passed back to the arbitrator.

COLIDE

The COLIDE portion of the topology module is called by the delay module when preparing to transmit a message. Input to COLIDE is provided by the variable present controlling processor (PP). Response is sent via the variable collision (CLSN using the node status array, NSTA). COLIDE checks the collision, indicator flag of the transmitting node. If the flag indicates a collision, the appropriate response is returned to the delay module and the involved nodes are taken off the bus and placed in an on-pending state. If no collision occurs, the appropriate response is returned via CLSN.

CNVRT

The Convert routine is used by the arbitrator to find the corresponding physical node number when given a logical address. The input for this routine is contained in the variable logical processor number (LPNO) and the corresponding physical node number is returned via the variable physical processor number (PPNO). Correlation is accomplished using the node status array (NSTA) attribute logical address. The present model assumes that any one physical node will have only one corresponding logical address.

RMVND

Removal of a node from the system is accomplished by using this topology call. The number of the node to be removed is passed via the variable process number (PRONO). The node status array (NSTA) active flag for that particular node is reset.

ADDND

The addition of a node to the system is accomplished using this routine and the node number passed via processor number (PRONO). The addition involves the assignment of a random number to the particular node and resets the bus time-out counter or the node status array.

DIST

This routine, which returns internodal distance for the transmission of a message, is called by the delay module. Its operation is functionally identical to the control distance routine (CDST) except that the destination of the message is found in the message attributes instead of next controlling processor (NP). All of the same routines are used here (ROUTE, SHRTE, TOTAL).

DVAL

Destination validity is checked by this routine, which is called by the delay module. It uses the destination found in the message attributes and checks the node status array (NSTA) for an active condition for that particular processor. It returns the result of the check via the variable valid destination (VALD).

BUSTO

The bus time-out processor provides a mechanism for updating sequence number and assigning new ones. It is called every time a bus time-out (BTO) appears in the system. The incrementing and checking of the node status array (NSTA) quantities by CCNT and RANU are done in this module. Also, upon assignment of a new sequence number, all others are checked for collision and the collision indicators are set.

As an overview, the topology module provides two major functions: (1) source to destination distance and routing statistics and (2) constant updating of the system topology model. Grouping the subroutines into these two functional categories yields (Table 10-5).

ANALYSIS MODULE

The goal of the Analysis Module is to (1) provide quantitative measure that establish the effectiveness of distributed processing systems; (2) provide statistical measures that can be used to compare distributed processing systems having divergent design philosophies. To meet these goals, it is necessary to identify constant factors that unify distributed processing systems and derive statistical measures by which these factors can be compared. That is to say, a "common

Table 10-5 Topology Subroutines Function and Classification

System Routine Statistics	Topology Updates
CSDT	RMVND
COLIDE	ADDND
CNVRT	BUSTO
DIST	TOPINIT*

TOPINIT is not physically contained in the topology module.

language" of analysis must be established by which a wide range of distributed systems can be described.

Establishing Analysis Criterion

In order to develop wide-ranging analysis criterion, it is necessary to identify those characteristics that are common among distributed processing systems. These common characteristics will be developed into statistical measures that analyze the relative merits of the underlying system. In developing common characteristics, three areas will be explored: (1) the basic physical structure of distributed processing networks, (2) the basic sequence of events, and (3) the overall function of distributed processing networks.

To provide flexibility and simplicity, most distributed processing systems have adopted a modular design philosophy. Modularity has resulted in a common physical structure that allows the distributed processing systems to be divided into several functional components. These component parts can be examined and evaluated separately. Dividing the evaluation of a system into functional components allows more accurate analysis of the intermediate factors that contribute to the strengths and weaknesses of a system.

The basic functional components that form the physical structure of a typical distributed processing system are shown in Figure 10-22. This diagram describes each distributed processing system in terms of the following components: (1) a number of processors that generate and consume messages, (2) a waiting line or queue containing messages that cannot be serviced immediately, (3) an interface unit that prepares messages for transmission, and (4) a communications network that performs the actual physical transmission of data. The physical implementation of these component parts differs widely from system to system. The outline presented in Figure 10-22 represents an accurate, generalized picture of distributed processing systems. The physical mapping presented in Figure 10-23 allows the identification of certain common features and checkpoints that are discussed in subsequent paragraphs.

The primary structural feature of quantitative interest in Figure 10-24 is the queue or waiting line. The length of these queues gives some quantitative information concerning the effectiveness of the underlying communication system. Exceptionally long or unbalanced queues could indicate the presence of system bottlenecks. Queues that grow and retreat wildly could suggest poor responsiveness to peak loads.

The basic components, which form the functional event structure in the typical distributed processing network, are shown in Figure 10-23. This figure reproduces the same general physical layout presented in Figure 10-22, but divides the passage of messages through the physical system into specific steps or phases. The major events of interest along the message path, Figure 10-23, are: (1) the message arrives, (2) the message enters the queue, (3) the message leaves the queue, (4) the message becomes available to the interface unit, (5) the message starts transmission, (6) the

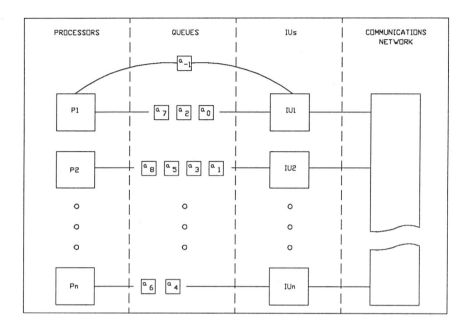

Figure 10-22 Functional components of a distributed processing system.

message ends transmission, and (7) the message becomes available to the receiving processor.

These common checkpoints are significant because they allow time measurements that chart the passage of the message through the system. As long as a particular system accurately implements communication, timing becomes a most critical factor. That is, the speed at which accurately transmitted messages are

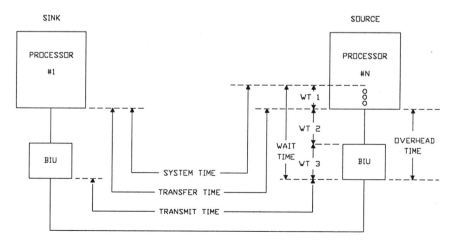

Figure 10-23 Bus evaluation parameters.

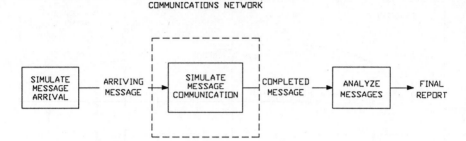

Figure 10-24 Simulated message flow graph.

completed is of primary interest. This series of checkpoints allows analysis of overall as well as intermediate delays imposed on the communication process.

The time between basic checkpoints and combinations of checkpoints gives rise to specific descriptive quantities, shown by the arrows in Figure 10-23. These quantities will be compiled for each simulation run on specific distributed processing networks. These are described in more detail, as follows:

1. System Time is the time between message generation at the source processor and message reception by the sink processor. System time quantifies the total delay the distributed processing system imposes on a message.
2. Transfer Time is the time that elapses between a message leaving the queue and its reception at the sink. Transfer time indicates the time required for the system to effect communications disregarding the time spent waiting to commence the transfer process.
3. Transmit Time is the time a message spends in the process of physical information transmission. This quantity indicates the actual timeliness of the low-level protocol and the speed of the physical transmission.
4. Wait Time is the time that must be expended before a message begins transmission. This quantity is divided into four smaller quantities: $wt1$, $wt2$, $wt3$, and $wt4$, described as follows:

    $wt1$ is the time a message spends in the queue.
    $wt2$ is the interval between removing the message from the queue to the point at which the IU begins preparing the message for transmission.
    $wt3$ accounts for the time required to prepare a message for transmission.
    $wt4$ includes the time required to make the message available to the sink processor once the transmission is complete.

5. Overhead Time equals the sum of $wt2$, $wt3$, and $wt4$. The overhead time is considered the time that must be yielded to the IU as the price of message transmission.

Analysis criterion can also be approached from a functional point of view. Functional criteria allows evaluation and comparison of the performance of distributed processing systems. These criterion fall into the following categories: 1) the amount of information carried by the communications system in unit time; i.e., throughput, 2) the amount of useful information transferred in unit time excluding overhead; i.e., information throughput, 3) the information lost in the communications process, 4) the amount of information overhead, and 5) the proportion of data that arrives late.

The throughput statistics quantify the total volume of information carried by a distributed processing system in unit time. This value is an overall indicator of the capacity and utilization of a distributed processing system. Unfortunately, the volume of "real" information transferred is reduced by the portion of overhead appended to the message. The overhead information is that part of a message that is attached by the distributed processing system to facilitate communication. The measurement of throughput that disregards overhead is called information throughput.

In addition to the overall flow of information through the system, we are interested in the loss of information. This loss can be the result of three conditions: (1) message loss because of a full queue, (2) message loss because of a bit(s) in error during transmission, (3) message loss because of the casualty of a system component. These quantities will be computed as a percent of the total number of messages transmitted. A statistic relating total messages lost will also be computed.

Also of interest is the amount of overhead that is attached to each message. This quantity allows analysis of the degradation of system performance caused by overhead. This statistic expresses overhead as a percentage of the total information transferred.

A more sophisticated functional evaluation criterion is the data late statistics. This statistic evaluates the proportion of messages that arrive past their time of expiration. In a very practical sense, this measure is one that is of ultimate concern in C3. If all messages arrive in their allotted time, the system is working within capacity and responds to the peak requirements demanded of it.

In summary, the criterion used for analysis is generated by three characteristics of distributed processing systems:

1. Basic physical structure. This characteristic yields criteria such as queue length, which result from the physical link between the processor and interface unit.
2. The event structure. This characteristic yields analysis criteria such as transmit time that is the result of the requirement to physically transmit the data during some point in the communication cycle.
3. The overall function. This characteristic yields analysis criteria such as throughput that result from the overall function of the system; i.e., to communicate.

Analysis Criteria and Simulation

From the point of view of statistical analysis, Figure 10-24 illustrates the essential nature of the model. The block on the left shows simulation messages arriving to the communication network from the C3 system. Messages in the simulation are not real system messages, but are buffers of computer memory that contain information regarding the nature and history of the message. The central block shows simulation of messages passing through the communication network. During this passage, data, which captures the history of the message, are added to the simulated message. When the message is either complete (reaches its destination) or lost (failed to reach its destination), the message is analyzed by the analysis module shown as the block on the right.

During a simulation, many messages will take the path illustrated in Figure 10-24. A large number of messages are required in order to build up what is called "statistical significance.This refers to the fact that a large sample of occurrences must be taken into consideration in order to eliminate any bias that may be produced by taking too small a random sample.

The structure of the software for collecting statistics and formatting the final report is shown in Figure 10-25. During a simulation run, information from large numbers of completed messages will be accumulated and stored, and the memory occupied by these messages will be released. At the end of a simulation run, these accumulated data will be used in statistical calculations, which will be formatted and presented in the form of a final report.

Statistical Output

The statistics generated by the system can be divided into three main groups: (1) time independent, (2) time persistent, and (3) periodic.

The time independent group are statistics that arise from independent observations. The traditional mean and standard deviation can be calculated for this group. These data, which are accumulated during the simulation run, are as follows: (1) the sum of each observed piece of data, (2) the sum of each piece of data squared, (3) the number of observations, and (4) the maximum value observed. From the accumulated data, the mean, standard deviation, and maximum observed value will be calculated and formatted for the final report. These statistics will be provided for all the time independent data points.

Figure 10-25 Analysis module functional block diagram.

Time persistent statistics are important when the time over which a parameter retains its value becomes critical. An example of this is a waiting line. If the line has 10 members in it for 20 minutes and 1 member for 1 minute, the average is not $(1 + 10)/2$, or 5.5. This quantity would indicate that there were approximately five members present in the line for a 21-minute period. The true average is more like $20/21 * 10 + 1/20 * 1$, or 9.57, or approximately 10. This is the time persistent average. As can be seen in this case, the average is weighted by the time period over which the value persisted. There is a similar argument that can be made for the time persistent standard deviation. These data, which are accumulated during a simulation for the time persistent case, are as follows: (1) the sum of the observed value times the period over which it retained that value, (2) the sum of the observed value squared times the period over which it retained its value, (3) the maximum observed value, and (4) the total period of observation. From these accumulated data, the time persistent mean, the time persistent standard deviation, and the maximum observed value will be calculated and formatted for the final report. These statistics will be provided for all the time persistent data points.

Periodic statistics are designed to yield a plot of observations as a function of time. This group of statistics affords a view of the system as it operates in time. Data are accumulated as in the previous two examples, except rather than sums of statistics, an individual data point graph of time versus the value of the data points will be plotted. These plots will be produced for all groups of periodic statistics.

The preceding sections have described the workings of the MALAN and general details about the analysis criterion and statistics that are drawn from these criteria (see Tables 10-6 and 10-7).

MALAN IMPLEMENTATION

The MALAN was developed in the spirit of providing a general-purpose, easily modifiable tool for the modeling of communication network protocols and topologies. The implementation philosophy was to modularize the network simulation software so that each module represented a major function found in a communication system. Within each module, a number of system implementation options can be selected to simulate the desired functional characteristics. Combinations of these functional modules are then pulled together to form a total system simulation model. All software modules are coded in FORTRAN and are extensively documented. If the need arises, therefore, a model developer can modify or add to existing modules to implement any desired characteristic that is not currently simulated. In this way, virtually any network protocol or topology can be simulated by choosing the desired system characteristics, defining the system topology, and, if necessary, implementing small sections of code for special purpose functions.

The network simulator currently consists of 73 modules and approximately 20K lines of FORTRAN code. A module typically has several subroutines that implement pieces of the total function and that perform simulation housekeeping.

Table 10-6 Simulator Events

1. Message arrivals
2. Message arbitration (control determination)
3. Message (un)blocking
4. Transfer for information (use event)
5. Reception of information (unfile event)
6. Statistical collection and reporting
7. Error determination and generation

Table 10-7 Scheduling of Events

1. Arrival module (process 1.0)
   a. Can schedule itself
   b. Can schedule Arbitrator (Request/Grant only)
   c. Calls analysis
2. Arbitrator module (process 2.0)
   a. Can schedule itself (daisy chain polling)
   b. Can schedule Use
   c. Calls Analysis
   d. Calls Topology
   e. Calls Block
3. (Un) blocking event module
   a. Calls blocking
   b. Can call Arbitrator
4. Transfer of information (use module) (process 3.0)
   a. Can schedule Unfile Event
   b. Calls Delay
   c. Calls Error
   d. Calls Topology
   e. Calls Analysis
5. Reception of information (Infile Event) (Process 3.5)
   a. Can schedule Arbitrator
   b. Calls analysis
   c. Calls topology.
6. Statistics module
   a. Scheduled by GASP
7. Error module (Processes 3.4 and 3.6)
   a. Calls Topology
   b. Calls analysis
8. The User control file
   a. Schedules simulation start
   b. Schedules end of simulation
   c. Schedules start of arrivals
   d. Schedules start of arbitration
   e. Schedules errors

Functionally, the simulator is broken up into four major sections: (1) topology, which simulates the physical layout and characteristics of the network; (2) arbitration, which simulates the control and access mechanisms for the physical medium (protocol simulation); (3) use, which simulates the actual use of the medium; and (4) analysis, which performs collection and calculation of simulation statistics and results. The following paragraphs describe each of the above sections. Figure 10-26 shows the general organization of the network simulator.

The topology for a particular system simulation is defined using a system definition program that allows interactive input of system parameters. The physical layout is defined in terms of communication links between nodes and in terms of the link speed and line length. Currently, the network simulator allows the definition and use of up to 10 communication busses. Each simulated bus may have its own specified protocol and transfer rate. Any node in the system may connect to any number of the simulated busses. This allows for the simulation of multiple interconnected and hybrid systems. During the simulation of a network, internodal messages are generated and distributed to the system nodes for transfer. The simulation topology function examines the source and destination node numbers associated with each message and determines a path by which the message should traverse the network. The path determination function is a replaceable algorithm that can be tailored to specific traffic routing strategies. The default routing strategy uses a "greedy" algorithm to find communication paths. Nodes that reside on the same communication bus, however, use the common medium for message transfer. For network simulations in which frequently communicating node resources are typically located on the same bus, this algorithm is quite sufficient.

Arbitration in the network simulator performs the actions associated with obtaining physical access to the communication bus. As simulated message traffic is generated, individual messages are assigned to nodes based on some user-defined distribution. The arbitrator judges which node will be given access to the data path based on the rules of the user-defined protocol. There are a number of predefined protocols in the network simulator from which a user can build a large number of different arbitration procedures. These include protocols for bus-, ring-, and star-type networks as well as time division multiplexed, daisy chained, round robin, and centralized control schemes. Also, user-defined protocols may be added or combined with existing types through the installation of user-supplied routines. In general, the simulation arbitrator controls the order, frequency, and duration of a node's access to the communication bus.

The use section of the simulator models the delays and actions associated with actual message transfer on the communication medium. Provisions are made for the user selection of message and bus parameters such as the number of retries for a failed message, the bit error rate of the bus, the data transfer speed of the bus, and the word size for transferred data. Use processing is mainly concerned with accumulating the delays that a message incurs during transfer. A secondary concern is the simulation of bus error conditions as specified by user input data for the bus reliability.

The network simulator analysis section collects data and parameters for the operational analysis of the modeled network. The information here is in addition to the data collection and analysis facilities provided by the GASP IV simulation language. Also, the analysis section uses some of the GASP IV collected data to calculate desired quantities such as system throughput, average system delay time, and the probability of data arriving late.

Using the four sections of the simulator described above, a typical network simulation operates as follows:

An initial system message load is generated to kick off the simulation.

At each message arrival (an event indicating that simulation time has advanced to a message arrival at a node), the message is placed in the node's communication queue.

For each new message arrival, a future arrival is generated and filed using the GASP IV event filing system.

The topology module is invoked to determine a message transfer path.

The bus arbitrator is invoked (a periodic event defined by the particular protocol being used) and decides which node will use the bus during this arbitration period.

The use of the communication medium is simulated for the message in question.

Message completion and analysis processing occurs.

These steps repeat until a predefined stopping condition is met. Typical stopping conditions are the passage of a prescribed amount of time or the transfer of a certain number of messages.

The major events simulated by MALAN are shown in Table 10-8. Their related interactions (possible events that can be scheduled) are shown in Table 10-9. Pseudo code for the major modules as well as a disk of the elements can be requested via the publisher.

REFERENCES

J.P. Buzen, "Computational Algorithms for Closed Networks with Exponential Servers," Communications of the ACM, Vol. 16, No. 9, 1978, pp. 527-531.

J.W. Boyse and D.R. Warn, "A Straightforward Model for Computer Performance Prediction," ACM Computing Surveys, June 1975, pp. 73-93.

A.A. Pritsker, The GASP IV Simulation Language, Wiley, New York, 1975.

P.J. Fortier, "A Communication Environment for Real-Time Distributed Control Systems, " Proceedings of the First ACM Northeast Regional Conference, 1984, pp. 371-382.

A. B. Pritsker, The GASP IV Simulation Language, John Wiley, 1974.

T. Schriber, "Simulation Using PPSS," John Wiley, 1974.

Kiviat P., Villanueva, R. Markowitz, The SIMSCRIPT II Programming Language, Prentice-Hall, 1969.

A.B. Pritsker, "Simulation and SLAM II," John Wiley and Son, 1984.

P.J. Fortier, Turner, P., A Simulation Program for Analysis of Distributerd Database Processing Concept. 19th Annual Simulation Symposium, 1986.

Table 10-8   IU Evaluation Parameters

I. Calculated For Each Processor

Queue Length Statistics

Average Queue Length	The time weighed average of the number of messages waiting for transfer
Peak Queue Length	The Maximum number of messages which have waited in the IU queue.
Standard Deviation of Queue Length	The standard deviation of the queue length from its average.

Message Size Statistics

Average Message Size	The average size of messages.
Maximum Message Size	The maximum size of a message.
Standard Deviation of Message Size	The standard deviation of the message size from its average.
Waiting Time Statistics	
Average Waiting Time	The average time a message is required to wait for service by the IU before it is transmitted.
Peak Waiting Time	The maximum time a message was required to wait before it was transmitted.
Standard Deviation of Waiting Time	The standard deviation of waiting time from its average.
Blocking Statistics	
Queue Block	The percentage of time arrivals are blocked by a full queue.
Contention Block	The percentage of time a queue is blocked due to a contention situation.

II Test Parameters Computed for the System

System Time Statistics

Average System Time	The average total time the message spends in the system. (i.e., the time it takes to reach a destination after it originates in the source)
Peak System Time	The maximum time a message spends in the transfer process.
Standard Deviation of System Time	The average deviation of system time from the average.

Table 10-8 IU Evaluation Parameters (cont.)

Messages in the System Statistics

Average Number of System Messages	Average number of messages contained in the system including those in transit.
Peak number of System Messages	The maximum number of messages contained in the system.
Standard Deviation of System Messages	The standard deviation of messages contained in the system from the average.
Transfer Time Statistics	
Average Message Transfer	The average time required to take messages Time from the origin and delivery it at its destination.
Peak Message Transfer Time	The maximum time a message takes to reach its destination.
Standard Deviation Time of Message Transfer	The standard deviation of message transfer times from their average.
Average Message Transmit Time	The average time a message spends in physical transmission over the communication lines.
Peak Message Transmit Time	The maximum time a message spends in physical transmission.
Standard Deviation Message Transmit Time	The standard deviation of the transmit of time from the average.

System Waiting Time Statistics

Average System Waiting Time	The average time waited by all messages prior to transmission.
Peak System Waiting Time	The maximum time any message has waited for Transmission.
Standard Deviation Time System Waiting	The standard deviation of message waiting of time for transmission from the average.

System Message Size Statistics

Average Message Size	The average size of all messages generated in the system in the system.
Maximum Message Size in the System	The largest message which has been generated by the system.
Standard Deviation of Message Sizes	The standard deviation of messages sizes from their average.

Table 10-8   IU Evaluation Parameters (cont.)

Overhead Time Statistics

Average Overhead Time	The average of message transfer time (message transmit time).
Peak Overhead Time	The maximum of message transfer time (message transmit time).
Standard Deviation of Overhead Time	The standard deviation of overhead time from its average.

Information Overhead Statistics

Average Information	
Average of Overhead	The total number of overhead bits ? the total number of message information bits.

Information Throughput Statistics

Average Information Throughput	The average number of information bits transferred over unit time.
Maximum Information Throughput	The maximum number of bits carried by the system in unit time.

System Throughput Statistics

Average System Throughput	The average number of bits transmitted in unit time.
Maximum System Throughput	The maximum number of bits carried by the system in unit time.

Message Lost Statistics

Total Messages Lost	The total number of messages lost.
Message Loss Rate	The rate of message loss.

Time to Complete Statistics

Probability of Data Late	The probability that a system message will arrive after it is set to expire.
Average Late Time	The average time messages were late in the system.

274 Modeling and Analysis of Local Area Networks

Table 10-8 IU Evaluation Parameters (cont.)

Message Loss Due to Full Queue — Messages lost because of queue overflow.

Message Loss Due to Bit Error (Percent) — Message loss because of errors in transmission.

Message Loss Due to Status Change (Percent) — Message lost due to loss of an IU or node.

III Periodic Tabular Data Collection

Periodic Average System Throughput Statistics

Periodic Average System Throughput — System throughput sampled periodically and displayed in graphic form.

Periodic Queue Length Statistics

Periodic System Queue Length — The sum of all processor queues sampled periodically and displayed in graph form.

Periodic Messages in the System Statistics

Periodic Measure of Messages System — The difference between: total messages in the generated and total messages received sampled periodically and displayed graphically.

Periodic Number of Retransmits

Periodic Retransmits — The number of retranmits which occur over a period of time and displayed graphically.

Table 10-9 Mathematical Presentation of Statistics

IU Statistics Computed for Each Processor

Queue Length Statistics

Average Queue Length (AQL) $\sum_{q} \frac{Q \times t(Q)}{Tot}$

Peak Queue Max Length (PQL) $\mathrm{Max} \Big|_{i=1}^{x} Q(i)$

Table 10-9 Mathematical Presentation of Statistics (cont.)

Standard Deviation of Queue Length $\quad \sqrt{\left(\sum_q Q^2 \times t(Q)\right) - (AQL)^2}$

Definition of Variables and Measurement Site

Q	The number of elements contained in a queue awaiting IU service. Value obtained from the Queue State Table contained in the Arrival Module.
$\text{Max} \begin{vmatrix} x \\ i=1 \end{vmatrix} Q(i)$	The maximum of all elements 1-x of Q Where x is the set of all unique values of Q
Tot	Total Simulation Time Value obtained from the simulation clock time TNOW at the moment of statistical calculation.
t(Q)	The time interval over which Q has retained its value. Value obtained from the time the statistic was last taken until TNOW.
Measurement Site	The Arrival Module
Analysis Site	The Analysis Module

Message Size Statistics

Average Message Size (AMS) $\quad \left(\sum_{i=1}^{m} MS(i)\right) / M$

Maximum Message Size $\quad \text{Max} \begin{vmatrix} m \\ i=1 \end{vmatrix} MS(i)$

Standard Deviation of Message Size $\quad \sqrt{\dfrac{\sum_{i=1}^{m}(MS(i) - AMS)^2}{M-1}}$

Definition of Variables and Measurement Site

M	Set of all messages in a processor
MS	The message size of any one of M messages in the system. Value obtained from Message Data Type, Attribute #7.

Table 10-9 Mathematical Presentation of Statistics (cont.)

Measurement/Analysis Site Analysis Module
$\text{Max}\Big|_{i=1}^{m} \text{MS}(i)$ The maximum message size of M Messages in the system.

Waiting Time Statistics

Average Waiting Time (AWT)

$$\left(\sum_{i=1}^{m} \text{WT}(i)\right)\bigg/ M$$

Maximum Waiting Time

$$\text{Max}\Big|_{i=1}^{m} \text{WT}(i)$$

Standard Deviation Waiting Time

$$\sqrt{\frac{\sum_{i=1}^{m} (\text{WT}(i) - \text{AWT})^2}{M-1}}$$

Definition of Variables and Measurement Site

M	Set of all messgaes in a processor.
WT	The Waiting Time of any one of M messages in the system.
Max WT	The maximum waiting time of M messages in the system.
Measurement/Analysis Site	Analysis Module

System Statistics Computed for the System

System Time Statistics

Average System Time (AST)

$$\sqrt{\frac{\sum_{i=1}^{SM} (\text{GT}(i) - \text{DAT}(i))}{SM}}$$

Peak System Time

$$\text{Max}\Big|_{i=1}^{SM} (\text{GT}(i) - \text{DAT}(i))$$

Standard Deviaton of System Time

$$\sqrt{\frac{\sum_{i=1}^{SM} ((\text{GT}(i) - \text{DAT}(i)) - \text{AST})^2}{SM}}$$

Table 10-9 Mathematical Presentation of Statistics (cont.)

Definition of Variables and Measurement Site

GT	Generation Time of the message obtained from message data type attribute #6.
GT	Measurement Site Arrival Module
DAT	Destination Arrival Time obtained from TNOW; the present simulated time at the time of arrival at the destination processor.
DAT	Measurement Site Analysis Module
SM	The set of all messages within the system.
$\text{Max} \left| \substack{SM \\ i=1} \right. (GT(i) - DAT(i))$	Computes the maximum of i-1 (GT(i)-DAT(i))
Analysis Site	Analysis Module

Messages in the System Statistics

Average Number of System Messages
$$\sqrt{\frac{\sum_{i=1}^{GR} (TMG - TMR) \times t(GRi)}{ToT}}$$

Peak Number of System Messages
$$\text{Max} \left| \substack{GR \\ i=1} \right. \text{left}(TMG - TMR) i$$

Standard Deviation of System Messages
$$\sum_{i=1}^{GR} ((TMG - TMR)^2 \times t(GRi)) - (AN - SM)^2$$

Definition of Variables and Measurement Sites

TMB	Total Messages Generated
TMG	Measurement Site Arrival Module
TMR	Total Messages Received (Completed)
TMR	Measurement Site Analysis Module
$t(GRi)$	The time between measurements (t_i and $t_i +1$)
ToT	Total Simulation Time
ToT	Measurement Site Analysis Module
$\text{Max} \left| \substack{GR \\ i=1} \right. (TMG - TMR) i$	Computes the maximum (TMG-TMR) i=1
Analysis Site	Analysis Module

278 Modeling and Analysis of Local Area Networks

Table 10-9 Mathematical Presentation of Statistics (cont.)

Transfer Time Statistics

Average Message Transfer Time $\qquad \sum_{i=1}^{SM} (XFER(i))/SM$

Peak Message Transfer Time $\qquad \text{Max} \begin{vmatrix} SM \\ i=1 \end{vmatrix} XFER(i)$

Standard Deviation of Message Transfer Time $\qquad \sqrt{\dfrac{\sum_{i=1}^{SM}(XFER(i) - AMTT)^2}{SM-1}}$

Definition of Variables and Measurement Sites

XFER	Message transfer time obtained from Message Data Type, Attribute #13.
SM	Set of all messages within the system.
XFER Measurement Site	Arrival Module and Delay Module.
$\text{Max} \begin{vmatrix} SM \\ i=1 \end{vmatrix} XFER(i)$	Calculates the maximum of XFER: i=1 XFER(i) -XFER(SM)
Analysis Site	Analysis Module

Transmit Statistics

Average Message Transmit Time (AMTR) $\qquad \sum_{i=1}^{SM} TT(i)/SM$

Peak Message Transmit Time $\qquad \text{Max} \begin{vmatrix} SM \\ i=1 \end{vmatrix} TT(i)$

Standard Deviation of Transmit Time $\qquad \sqrt{\dfrac{\sum_{i=1}^{SM}(TT(i) - AMTR)^2}{SM-1}}$

Definition of Variables and Measurement Site

TT	Message Transmit Time Obtained from a Message Data Type Attribute #12
TT Measurement Site	Delay Module
SM	The set of all messages in the system
Analysis Site	Analysis Module

Table 10-9 Mathematical Presentation of Statistics (cont.)

$\text{Max} \left| \sum_{i=1}^{SM} TT(i) \right|$ Calculate the maximum transmit i=1 time for the system.

System Waiting Time Statistics

Average System Waiting Time (ASWT) $\left(\sum_{i=1}^{SM} WT(i) \right) / SM$

Peak System Waiting Time $\text{Max} \left| \sum_{i=1}^{SM} WT(i) \right|$

Standard Deviation of System Wait Time $\sqrt{ \sum_{i=1}^{SM} (WT(i) - ASWT)^2 / SM - 1 }$

Definition of Variables and Measurement Site

WT	The waiting time of system messages.
SM	Set of all messages in the system.
WT = WT1 + WT2 + WT3	
WT1	Time message waits in the Queue.
Measurement Site WT1	Arrival Module
WT2	Time between leaving the queue and entering the BIU.
Measurement Site WT2	Use Module
WT3	The time required by the BIU to start transmission.
Measurement Site WT3	Use Module
$\text{Max} \left\vert \sum_{i=1}^{SM} WT(i) \right\rbrace$	Computes the maximum waiting time i=1 of any message in the system.

System Message Size Statistics

Average System Message Size (ASMS) $\left(\sum_{i=1}^{SM} MS(i) \right) / SM$

Peak System Message Size $\text{Max} \left| \sum_{i=1}^{SM} MS(i) \right|$

Standard Deviation of System Message Size $\sqrt{ \sum_{i=1}^{SM} (MS(i) - ASMS)^2 / SM - 1 }$

Table 10-9 Mathematical Presentation of Statistics (cont.)

Definition of Variables and Measurement Site

MS	Message Size of any one of SM messages in the system obtained from Message Data Type Attribute #7.
MS	Measurement Site Analysis Module
SM	The set of all messages in the system.
$\text{Max} \begin{vmatrix} MS \\ i=1 \end{vmatrix} MS(i)$	Calculates the maximum message i=1 size of all messages in the system.
Analysis Site	Analysis Module

Overhead Time Statistics

Average Overhead Time (AOT)

$$\left(\sum_{i=1}^{SM} XFER(i) - TT(i) \right) / SM$$

Peak Overhead Time

$$\text{Max} \begin{vmatrix} SM \\ i=1 \end{vmatrix} (XFER(i) - TT(i))$$

Standard Deviation of Overhead Time

$$\sum_{i=1}^{SM} ((XFER(i) - TT(i)) - AOT)^2 / SM - 1$$

Definition of Variables and Measurement Site

XFER	Message transfer time obtained from message data type attribute #13
XFER Measurement Site	Arrival Module and Delay Module
TT	Message Transmit Time
TT Measurement Site	Use Module
SM	Set of all messages within the system
$\text{Max} \begin{vmatrix} SM \\ i=1 \end{vmatrix} (XFER(i) - TT(i))$	Calculates the maximum of i=1 (XFER(i)- TT(i)
Analysis Site	Analysis Module

Information Overhead Statistics

Average Information Overhead

$$\left(\sum_{i=1}^{SM} MO(i) \ 8 \times MS(i) \right) / SM$$

Table 10-9 Mathematical Presentation of Statistics (cont.)

Definition of Variables and Measurement Sites

MO	Message Overhead (Bits) obtained from Message Data Type Attribute #8.
Measurement Site	Use Module
SM	Set of all messages in the system.
MS	Message Size obtained
Analysis Site	Analysis Module.

Information Throughput Statistics

Average Information Throughput
$$\left(\sum_{i=1}^{SM} MO(i)\ 8 \times MS(i) \right) / SM$$

Maximum Information Throughput
$$\text{Max} \left|_{i=1}^{MS} MS(i) \times 8/t(i) \right.$$

Definition of Variables and Measurement Sites

MS	Message Size of any one of SM messages in the system.
MS Measurement Site	Analysis Module.
SM	The set of all messages in the system.
t(i)	Time over which the number of messages in the system is constant
$\text{Max}\left\|_{i=1}^{SM} MS(i) \times 8/t(i)\right.$	The maximum value the Information Throughput has attained
Analysis Site	Analysis Module

System Throughput Statistics

Average System Throughput
$$\left(\sum_{i=1}^{SM} \frac{MS(i) \times 8 + MO(i)}{t(i)} \right) / SM$$

Maximum System Throughput
$$\text{Max} \left|_{i=1}^{MS} \frac{MS(i) \times 8 + MO(i)}{t(i)} \right.$$

Table 10-9 Mathematical Presentation of Statistics (cont.)

Definition of Variables and Measurement Sites

MS	Message size of any one of SM messages in the system obtained from message data type attribute #7.
MS Measurement Site	Analysis Module
SM	The set of all messages in the system.
t(i)	Time over which the number of messages in the system if constant.
MO	Message Overhead (bits) obtained from message data type attribute #8.
Analysis Site	Analysis Module.

Message Loss Statistics

Total Messages Lost (TML)
$$\sum_{i=1}^{SM} ML(i)$$

Message Lost Rate   TML/TOT

Message Lost due to Full Queue (FQ)
$$\left(\sum_{i=1}^{SM} EQU(1, ML)\right) / TML \times 100\%$$

Message Lost due to Bit Error (BE)
$$\left(\sum_{i=1}^{SM} EQU(2, ML)\right) / TML \times 100\%$$

Message Lost due to Status Change (SC)
$$\left(\sum_{i=1}^{SM} EQU(3, ML)\right) / TML \times 100\%$$

Definitions of Variables and Measurement Sites

ML	Messages Lost by the system =0 - Not Lost =1 - Lost due to full queue =2 - Lost due to bit error =3 - Lost due to status change
Measurement Site	Error Module
SM	The set of all messages in the system.
TOT	The total time of a simulation run.

Table 10-9 Mathematical Presentation of Statistics (cont.)

Analysis Site   Analysis Module
EQU (N,ML)   Function which yields a "1" when (N=ML).

Time to Complete Statistics

Probability of Data Late (PDL)

$$\sum_{i=1}^{SM} \left(\frac{(ABS\ (TNOW(i) - MTTC(i)) - (TNOW(i) - MTTC(i))}{2(TNOW(i) - MTTC)} \right) \times \frac{1}{SM}$$

Average Late Time

$$\frac{\sum_{i=1}^{SM} \left(\frac{(ABS\ (TNOW(i)) - MTTC(i))}{2(TNOW(i) - MTTC(i))} \times MTTC - TNOW \right)}{PDL \times SM}$$

Definitions of Variables and Measurement Sites

MTTC	Message Time to Complete — time deadline of a message.
TNOW	The present time.
SM	Set of all messages in the system.
Measurement Site	Arrival Module
Analysis Site	Analysis Module

Periodic Tabular Data Collection

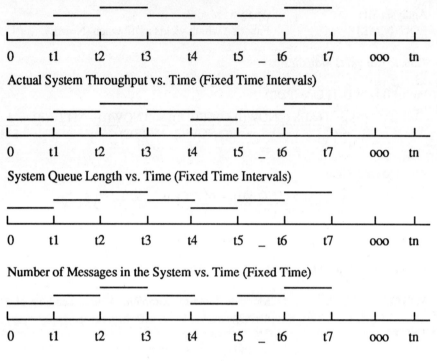

Appendix A

HXDP Calculations

Calculation of values found in graphs

$\lambda = 1/100$ milliseconds = 10 arrivals/second
$\lambda = 1$/second
$\lambda = 1/10$ seconds

Formula for average scan time

$$\bar{\tau} = \frac{\sum_{i=1}^{N} |i - (i+1)| t_R}{(1 - \frac{(t_s + T_{ack})}{N} (\sum_{k=1}^{N} (k^2 + (1+N)(-k + \frac{1}{2}N)) \cdot \tau_k)}$$

For simplicity, it was assumed that λk, $k=1,...,N$ were equal to λ, or

$$1 - \frac{(T_s + T_{ack}) \cdot 1)}{N} (\sum_{k=1}^{N} (k^2 + (1-n)(-k + \frac{1}{2}N)))$$

which simplifies to:

$$\bar{\tau} = \frac{\sum_{i=1}^{N} |i - (i+1)| t_R}{1 - \frac{t_s + T_{ack}}{N} \frac{(N(N+1)(2N+1)}{N} - \frac{(1+N)(N)(N+1)}{2} \frac{(1+N) N^2}{2}}$$

Where i is the logical number of the interface unit.

Configuration I computations: 16 processors

Given: Separation from IU 1 to IU 16 = 450 meters. Distance, d, therefore, equals 30 meters. The reallocation signal is equivalent to 6 bits. The ACK has 6 bits. The speed of a bit = 39.37 x 10^{-9} seconds/meter. The number of overhead bits/message = 71 bits.

Calculate:

$$\sum_{i=1}^{N} |i-(i+1)|, \text{ where } N+1 \equiv 1; = 30$$

Formula (1) becomes:

$$\bar{\tau} = \frac{30 \times 6 \text{ bits} \times 30 \text{ meters} \times 39.37 \times 10^{-9}/\text{meter} - \text{bit}}{1 - \frac{(t_s + t_{ack}) \times 1360}{16}}$$

$$\bar{\tau} = \frac{212.598 \times 10^{-6} \text{ seconds}}{1 - 85 \times (t_s + t_{ack}) \times \lambda}$$

$$t_s = \text{\# of bits} \times 39.37 \times 30 \times 10^{-9} \text{ seconds}$$
$$t_{sck} = 6 \times 39.37 \times 39 \times 10^{-9} \text{ seconds}$$

The values for λ = 10/second, 1/second, and 1/10 second are then obtained for messages ranging from 100 bits (100 bits + 71 overhead bits) to 800 bits.

Configuration II computations: 16 processors

Given: The separation from IU 1 to IU 16 = 450 meters. Distance, d, therefore equals 30 meters. The reallocation signal is equivalent to 6 bits. The ACK has 6 bits. The speed of a bit = 39.37 x 10^{-9} seconds/meter. The number of overhead bits/message = 71 bits.

Calculate:

$$\sum_{i=1}^{N} |i-(i+1)|, \text{ where } N+1 \equiv 1; = 128$$

Equation (1) becomes

$$\overline{\tau} = \frac{128 \times 6 \text{ bits} \times 30 \text{ meters} \times 39.37 \times 10^{-9}/\text{meter} - \text{bit}}{1 - 85 \times (t_s + t_{ack}) \times \lambda}$$

t_s = number of bits $\times$ 39.37 $\times$ 30 $\times$ 10^{-9} seconds
t_{ack} = 6 $\times$ 39.27 $\times$ 30 $\times$ 10^{-9} seconds

$$\overline{\tau} = \frac{.907 \times 10^{-3} \text{ seconds}}{1 - 85 \times (t_s + t_{ack}) \times \lambda}$$

The values for λ = 10/second, 1/second, and 1/10 second are then obtained for messages ranging from 100 to 800 bits.

Configuration I computations: 64 processors

Given: Separation from IU 1 to IU 64 = 450 meters. The distance, d, therefore equals 7.14 meters. The reallocation signal is equivalent to 6 bits. The ACK has 6 bits. The speed of a bit = 39.37 x 10-9 seconds/meter. The number of overhead bits/message = 71 bits.

Calculate:

$$\sum_{i=1}^{N} |i - (i+1)|, \text{ where } N + 1 \equiv 1; = 126$$

Equation (1) becomes

$$\overline{\tau} = \frac{126 \times 6 \text{ bits} \times 7.14 \text{ meters} \times 39.37 \times 10^{-9} \text{ meters/bit} - \text{sec}}{1 - 1365 \times (t_s + t_{ack}) \times \lambda}$$

t_s = number of bits $\times$ 39.37 $\times$ 7.14 $\times$ 10^{-9} seconds
t_{ack} = 6 $\times$ 39.37 $\times$ 7.14 $\times$ 10^{-9} seconds

$$\overline{\tau} = \frac{212.512 \times 10^{-6}}{1 - 1365 \times (t_s + t_{ack}) \times \lambda}$$

The values for λ = 10/seconds, 1/seconds, and 1/10 second are then obtained for messages ranging from 100 to 800 bits.

Configuration II computations: 64 processors

Given: The separation from IU 1 to IU 64 = 450 meters. Distance, d, therefore equals 7.14 meters. The reallocation signal is equivalent to 6 bits. The

288 Modeling and Analysis of Local Area Networks

ACK has 6 bits. The speed of a bit = 39.37 x 10-9 seconds/meter. The number of overhead bits/message = 71 bits.

Calculate:

$$\sum_{i=1}^{N} |i - (i+1)|, \text{ where } N + 1 \equiv 1; = 2048$$

Equations (1) becomes:

$$\bar{\tau} = \frac{2048 \times 6 \text{ bits} \times 7.14 \text{ meters} \times 39.37 \times 10^{-9} \text{ meters/bit} - \text{sec}}{1 - 1365 \times (t_s + t_{ack}) \times \lambda}$$

$$t_s = \text{number of bits} \times 39.37 \times 7.14 \times 10^{-9} \text{ seconds}$$
$$t_{ack} = 6 \times 39.37 \times 7.14 \times 10^{-9} \text{ seconds}$$

$$\bar{\tau} = \frac{3.454 \times 10^{-3}}{1 - 1365 \times (t_s + t_{ack}) \times \lambda}$$

The values for ? = 1/second and 1/10 second are then obtained for messages from 100 to 800 bits.

Configuration I calculations: 16 processors

$\lambda = 10/\text{second}$

1. bits = 171; $\bar{\tau} = \dfrac{212.598 \times 10^{-6}}{1 - 85 \times 10 \times (708.48 \times 10^{-9} + 201,916.8 \times 10^{-9})}$

   $\bar{\tau} = \dfrac{212.598 \times 10^{-6}}{1 - 850 \times 202.625 \times 10^{-6}}$

   $\bar{\tau} = \dfrac{212.598 \times 10^{-6}}{1 - .17223}$

   $= \dfrac{212.598 \times 10^{-6}}{.82777} = .2568 \times 10^{-3}$

2. bits = 271; $\bar{\tau} = \dfrac{212.598 \times 10^{-6}}{1 - 850 \times (708.48 \times 10^{-9} + 319 \times 996 \times 10^{-9})}$

   $\bar{\tau} = \dfrac{212.598 \times 10^{-6}}{1 - 850 \times (320.705 \times 10^{-6})}$

$$\bar{\tau} = \frac{212.598 \times 10^{-6}}{1 - .2725}$$

$$= \frac{212.598 \times 10^{-6}}{.7275} = .2922 \times 10^{-3}$$

3. bits = 471; $\bar{\tau} = \dfrac{212.598 \times 10^{-6}}{1 - 850 \times (708.48 \times 10^{-9} + 556{,}156.8 \times 10^{-9})}$

$$\bar{\tau} = \frac{212.598 \times 10^{-6}}{1 - 850 \times (556.865 \times 10^{-6})}$$

$$\bar{\tau} = \frac{212.598 \times 10^{-6}}{1 - .4733}$$

$$= \frac{212.598 \times 10^{-6}}{.5267} = .4036 \times 10^{-3}$$

4. bits = 641; $\bar{\tau} = \dfrac{212.598 \times 10^{-6}}{1 - 850 \times (708.48 \times 10^{-9} + 792{,}316.8 \times 10^{-9})}$

$$\bar{\tau} = \frac{212.598 \times 10^{-6}}{1 - 850 \times (793.025 \times 10^{-6})}$$

$$\bar{\tau} = \frac{212.598 \times 10^{-6}}{1 - .6740}$$

$$= \frac{212.598 \times 10^{-6}}{.326} = .6521 \times 10^{-3}$$

5. bits = 871; $\bar{\tau} = \dfrac{212.598 \times 10^{-6}}{1 - 850 \times (708.48 \times 10^{-9} + 1{,}028{,}476.8 \times 10^{-9})}$

$$\bar{\tau} = \frac{212.598 \times 10^{-6}}{1 - 850 \times (1{,}029{,}185.2 \times 10^{-9})}$$

$$\bar{\tau} = \frac{212.598 \times 10^{-6}}{1 - .874807}$$

$$= \frac{212.598 \times 10^{-6}}{.1251} = 1.699 \times 10^{-3}$$

$\lambda = 1/\text{second}$

1. bits = 171; $\bar{\tau} = \dfrac{212.598 \times 10^{-6}}{1 - 85 \times 202.625 \times 10^{-6}}$

$$\bar{\tau} = \frac{212.598 \times 10^{-6}}{1 - .017223}$$

$$= \frac{212.598 \times 10^{-6}}{.9827} = .2163 \times 10^{-3}$$

2. bits = 271; $\bar{\tau} = \dfrac{212.598 \times 10^{-6}}{1 - 85 \times (320.705 \times 10^{-6})}$

$\bar{\tau} = \dfrac{212.598 \times 10^{-6}}{1 - .02725}$

$\bar{\tau} = \dfrac{212.598 \times 10^{-6}}{.9727}$

$= .2185 \times 10^{-3}$ seconds

3. bits = 471; $\bar{\tau} = \dfrac{212.598 \times 10^{-6}}{1 - 85 \times (556.865 \times 10^{-6})}$

$\bar{\tau} = \dfrac{212.598 \times 10^{-6}}{1 - .04733}$

$\bar{\tau} = \dfrac{212.598 \times 10^{-6}}{.9526}$

$= .2331 \times 10^{-3}$ seconds

4. bits = 671; $\bar{\tau} = \dfrac{212.598 \times 10^{-6}}{1 - 85 \times (793.025 \times 10^{-6})}$

$\bar{\tau} = \dfrac{212.598 \times 10^{-6}}{1 - .0674}$

$\bar{\tau} = \dfrac{212.598 \times 10^{-6}}{.9326}$

$= .2279 \times 10^{-3}$ seconds

5. bits = 871; $\bar{\tau} = \dfrac{212.598 \times 10^{-6}}{1 - 85 (1{,}029{,}185.2 \times 10^{-9})}$

$\bar{\tau} = \dfrac{212.598 \times 10^{-6}}{1 - .08748}$

$\bar{\tau} = \dfrac{212.598 \times 10^{-6}}{.9125}$

$= .2329 \times 10^{-3}$ seconds

$\lambda = 1/10$ second

1. bits = 171; $\bar{\tau} = \dfrac{212.598 \times 10^{-6}}{1 - 85 \times 202.625 \times 10^{-9}}$

$\bar{\tau} = \dfrac{212.598 \times 10^{-6}}{1 - .0017223}$

$\bar{\tau} = \dfrac{212.598 \times 10^{-6}}{.9982}$

$= .2129 \times 10^{-3}$ seconds

2. bits = 271; $\bar{\tau} = \dfrac{212.598 \times 10^{-6}}{1 - 8.5 \times (320.705 \times 10^{-9})}$

$\bar{\tau} = \dfrac{212.598 \times 10^{-6}}{1 - .002725}$

$\bar{\tau} = \dfrac{212.598 \times 10^{-6}}{.9972}$

$= .2131 \times 10^{-3}$ seconds

3. bits = 471; $\bar{\tau} = \dfrac{212.598 \times 10^{-6}}{1 - 8.5 \times (556.865 \times 10^{-6})}$

$\bar{\tau} = \dfrac{212.598 \times 10^{-6}}{1 - .004733}$

$\bar{\tau} = \dfrac{212.598 \times 10^{-6}}{.9952}$

$= .2136 \times 10^{-3}$ seconds

4. bits = 671; $\bar{\tau} = \dfrac{212.598 \times 10^{-6}}{1 - 8.5 \times (793.025 \times 10^{-6})}$

$\bar{\tau} = \dfrac{212.598 \times 10^{-6}}{1 - .00674}$

$\bar{\tau} = \dfrac{212.598 \times 10^{-6}}{.9932}$

$= .2140 \times 10^{-3}$ seconds

5. bits = 871; $\bar{\tau} = \dfrac{212.598 \times 10^{-6}}{1 - 8.5 \times (1{,}029{,}185.2 \times 10^{-9})}$

$\bar{\tau} = \dfrac{212.598 \times 10^{-6}}{1 - .008748}$

$\bar{\tau} = \dfrac{212.598 \times 10^{-6}}{.9912}$

$= .2144 \times 10^{-3}$ seconds

Configuration II calculations: 16 processors

$\lambda = 10$/second

1. bits - 171; $\bar{\tau} = \dfrac{.907 \times 10^{-3}}{1 - 850 \times 202.625 \times 10^{-6})}$

$\bar{\tau} = \dfrac{.907 \times 10^{-3}}{.82777}$

$\bar{\tau} = 1.09 \times 10^{-3}$

2. bits = 271; $\bar{\tau} = \dfrac{.907 \times 10^{-3}}{1 - 850 \times (320.705 \times 10^{-6})}$

$\bar{\tau} = \dfrac{.907 \times 10^{-3}}{.7275}$

$= 1.246 \times 10^{-3}$ seconds

3. bits = 471; $\bar{\tau} = \dfrac{.907 \times 10^{-3}}{1 - 850 \times (556.865 \times 10^{-6})}$

$\bar{\tau} = \dfrac{.907 \times 10^{-3}}{.5267}$

$= 1.722 \times 10^{-3}$ seconds

4. bits = 671; $\bar{\tau} = \dfrac{.907 \times 10^{-3}}{1 - 850 \times (793.025 \times 10^{-6})}$

$\bar{\tau} = \dfrac{.907 \times 10^{-3}}{.326}$

$= 2.78 \times 10^{-3}$ seconds

5. bits = 871; $\bar{\tau} = \dfrac{.907 \times 10^{-3}}{.1251}$

$= 7.25 \times 10^{-3}$ seconds

$\lambda = 1/\text{second}$

1. bits = 171; $\bar{\tau} = \dfrac{.907 \times 10^{-3}}{.9827} = .9229 \times 10^{-3}$ seconds

2. bits = 271; $\bar{\tau} = \dfrac{.907 \times 10^{-3}}{.9727} = .9324 \times 10^{-3}$ seconds

3. bits = 471; $\bar{\tau} = \dfrac{.907 \times 10^{-3}}{.9526} = .9521 \times 10^{-3}$ seconds

4. bits 671; $\bar{\tau} = \dfrac{.907 \times 10^{-3}}{.9326} = .9725 \times 10^{-3}$ seconds

5. bits =871; $\bar{\tau} = \dfrac{.907 \times 10^{-3}}{.9125} = .9939 \times 10^{-3}$ seconds

$\lambda = 1/10$ second

1. bits =171; $\bar{\tau} = \dfrac{.907 \times 10^{-3}}{.9982} = .908 \times 10^{-3}$ seconds

2. bits = 271; $\bar{\tau} = \dfrac{.907 \times 10^{-3}}{.9972} = .9095 \times 10^{-3}$ seconds

3. bits = 471; $\bar{\tau} = \dfrac{.907 \times 10^{-3}}{.9952} = .9113 \times 10^{-3}$ seconds

4. bits = 671; $\bar{\tau} = \dfrac{.907 \times 10^{-3}}{.9932} = .9132 \times 10^{-3}$ seconds

5. bits = 871; $\bar{\tau} = \dfrac{.907 \times 10^{-3}}{.9912} = .9150 \times 10^{-3}$ seconds

Configuration I calculations: 64 processors

$\lambda = 1/\text{second}$

1. bits = 171; $\bar{\tau} = \dfrac{212.12 \times 10^{-6}}{1 - 1365 \times (1686.18 \times 10^{-9} + 48{,}056.19 \times 10^{-9})}$

    $\bar{\tau} = \dfrac{212.512 \times 10^{-6}}{1 - 1365 \times 49.742 \times 10^{-6})}$

    $\bar{\tau} = \dfrac{212.512 \times 10^{-6}}{1 - .06789} = \dfrac{212.512 \times 10^{-6}}{.9321}$

    $= .2279 \times 10^{-3}$ seconds

2. bits = 271; $\bar{\tau} = \dfrac{212.12 \times 10^{-6}}{1 - 1365 \times (1686.18 \times 10^{-9} + 48{,}056.19 \times 10^{-9})}$

    $\bar{\tau} = \dfrac{212.512 \times 10^{-6}}{1 - 1365 \times (77{,}845.5 \times 10^{-6})}$

    $\bar{\tau} = \dfrac{212.512 \times 10^{-6}}{1 - .106258} = \dfrac{212.512 \times 10^{-6}}{.8937}$

    $= .2377 \times 10^{-3}$ seconds

3. bits = 471; $\bar{\tau} = \dfrac{212.12 \times 10^{-6}}{1 - 1365 \times (1686.18 \times 10^{-9} + 132{,}365.31 \times 10^{-9})}$

    $\bar{\tau} = \dfrac{212.512 \times 10^{-6}}{1 - 1365 \times (134.051 \times 10^{-6})}$

    $\bar{\tau} = \dfrac{212.512 \times 10^{-6}}{1 - .1829}$

    $\bar{\tau} = \dfrac{212.512 \times 10^{-6}}{.8171} = .260 \times 10^{-3}$

4. bits = 671; $\bar{\tau} = \dfrac{212.12 \times 10^{-6}}{1 - 1365 \times (1686.18 \times 10^{-9} + 188{,}571.39 \times 10^{-9})}$

    $\bar{\tau} = \dfrac{212.512 \times 10^{-6}}{1 - 1365 \ (190.257 \times 10^{-6})}$

294   Modeling and Analysis of Local Area Networks

$$\bar{\tau} = \frac{212.512 \times 10^{-6}}{1 - .2597}$$

$$= \frac{212.512 \times 10^{-6}}{.7403} = .287 \times 10^{-3}$$

5. bits = 871; $\bar{\tau} = \dfrac{212.12 \times 10^{-6}}{1 - 1365 \times (1686.18 \times 10^{-9} + 244{,}777.47 \times 10^{-9})}$

$$\bar{\tau} = \frac{212.512 \times 10^{-6}}{1 - 1365 \times (246.463 \times 10^{-6})}$$

$$\bar{\tau} = \frac{212.512 \times 10^{-6}}{1 - .3364}$$

$$\bar{\tau} = \frac{212.512 \times 10^{-6}}{.6636}$$

$$= .3202 \times 10^{-3} \text{ seconds}$$

$\lambda = 1/10$ second

1. bits = 171; $\bar{\tau} = \dfrac{212.512 \times 10^{-6}}{1 - .006789}$

$$\bar{\tau} = \frac{212.512 \times 10^{-6}}{.9932} = .2139 \times 10^{-3} \text{ sec}$$

2. bits = 271; $\bar{\tau} = \dfrac{212.512 \times 10^{-6}}{1 - .0106}$

$$\bar{\tau} = \frac{212.512 \times 10^{-6}}{.9894} = .2147 \times 10^{-3} \text{ sec}$$

3. bits = 471; $\bar{\tau} = \dfrac{212.512 \times 10^{-6}}{1 - .01829}$

$$\bar{\tau} = \frac{212.512 \times 10^{-6}}{.9817} = .2164 \times 10^{-3} \text{ sec}$$

4. bits = 671; $\bar{\tau} = \dfrac{212.512 \times 10^{-6}}{1 - .02597}$

$$\bar{\tau} = \frac{212.512 \times 10^{-6}}{.974} = .2181 \times 10^{-3} \text{ sec}$$

5. bits = 871; $\bar{\tau} = \dfrac{212.512 \times 10^{-6}}{1 - .03364}$

$$\bar{\tau} = \frac{212.512 \times 10^{-6}}{.9663} = .2190 \times 10^{-3} \text{ sec}$$

Configuration II calculations : 64 processors

$\lambda = 1/\text{second}$

1. bits = 171; $\bar{\tau} = \dfrac{3.454 \times 10^{-3}}{.93210}$
   $\bar{\tau} = 3.705 \times 10^{-3}$ seconds

2. bits = 271; $\bar{\tau} = \dfrac{3.454 \times 10^{-3}}{.8937}$
   $\bar{\tau} = 3.86 \times 10^{-3}$ seconds

3. bits = 471; $\bar{\tau} = \dfrac{3.454 \times 10^{-3}}{.8171}$
   $\bar{\tau} = 4.227 \times 10^{-3}$ seconds

4. bits = 671; $\bar{\tau} = \dfrac{3.454 \times 10^{-3}}{.7403}$
   $\bar{\tau} = 4.665 \times 10^{-3}$ seconds

5. bits = 871; $\bar{\tau} = \dfrac{3.454 \times 10^{-3}}{.6636}$
   $\bar{\tau} = 5.204 \times 10^{-3}$ seconds

$\lambda = 1/10$ second

1. bits = 171; $\bar{\tau} = \dfrac{3.454 \times 10^{-3}}{.9932}$
   $\bar{\tau} = 3.477 \times 10^{-3}$ seconds

2. bits = 271; $\bar{\tau} = \dfrac{3.454 \times 10^{-3}}{.9894}$
   $\bar{\tau} = 3.491 \times 10^{-3}$ seconds

3. bits = 471; $\bar{\tau} = \dfrac{3.454 \times 10^{-3}}{.9817}$
   $\bar{\tau} = 3.518 \times 10^{-3}$ seconds

4. bits = 671; $\bar{\tau} = \dfrac{3.454 \times 10^{-3}}{.974}$
   $\bar{\tau} = 3.546 \times 10^{-3}$ seconds

5. bits = 871; $\bar{\tau} = \dfrac{3.454 \times 10^{-3}}{.9663}$
   $\bar{\tau} = 3.57 \times 10^{-3}$ seconds

Appendix B

Token Bus Computations

Comment: The following is from the design document for the Hughes IR&D bus Calculation of values found in graphs arrival rates are

$\lambda = 1/100$ milliseconds $= 10$ arrivals/second
$\lambda = 1$ second
$\lambda = 1/10$ seconds

The equations for average scan time is

$$\bar{\tau} = \frac{\sum_{i=1}^{N} |i - (i+1)| t_c}{(1 - \frac{t_s}{N-1}(\sum_{i=1}^{N} |k^2 + (1+N)(-k+\frac{1}{2}N)) \cdot \lambda_k)}$$

For purposes of simplicity, it was assumed that $\lambda_k, k=1,...N$ were equal to λ, or

$$\bar{\tau} = \frac{\sum_{i=1}^{N} |i - (i+1)| t_c}{1 - \frac{t_s \cdot \lambda}{N-1}(\sum_{i=1}^{N} |k^2 + (1+N)(-k+\frac{1}{2}N))}$$

which simplifies to:

298 Modeling and Analysis of Local Area Networks

$$(1) \quad \bar{\tau} = \frac{\sum_{i=1}^{N} |i-(i+1)| t_c}{1 - \frac{t_s \cdot \lambda}{N-1} \left(\frac{N \cdot (N+1)(2N+1)}{6} - \frac{(1+N)(N)(N+1)}{2} + \frac{(1+N)N^2}{2} \right)}$$

where i is the logical number of the interface unit.

Configuration I computations: 16 processors

Given: The separation from IU 1 to IU 16 = 450 meters. Distance, d, therefore equals 30 meters. A token has 21 bits. The speed of a bit = 1×10^{-9} seconds/inch or 39.37×10^{-9} seconds/meter. The number of overhead bits/message = 42 bits.

Calculate;

$$\sum_{i=1}^{N} |i-(i+1)|, \text{ where } N+1 \equiv 1; = 30$$

Equation (1) becomes

$$\bar{\tau} = \frac{30 \times 21 \text{ bits} \times 30 \text{ meters} \times 39.37 \times 10^{-9}/\text{meter} - \text{bit}}{1 - \frac{t_s \times 1360 \times \lambda}{15}}$$

$$\bar{\tau} = \frac{744.093 \times 10^{-6} \text{ seconds}}{1 = 90.66 \times t_s \times \lambda} \quad t_s = \text{number of bits} \times 39.36 \times 30 \times 10^{-9} \text{ seconds}$$

The values for λ = 10/second, 1/second, and 1/10 second are then obtained for messages ranging from 100 bits (100 bits + 42 overhead bits) to 800 bits.

Configuration II computations: 16 processors

Given: The separation from IU 1 to IU 16 = 450 meters. Distance, d, therefore equals 30 meters. A token has 21 bits. The speed of a bit = 1×10^{-9} seconds/inch or 39.37×10^{-9} seconds/meter. The number of overhead bits/message = 42 bits.

Calculate:

$$\sum_{i=1}^{N} |i - (i+1)|, \text{ where } N + 1 \equiv 1; = 128$$

Equation (1) becomes:

$$\bar{\tau} = \frac{128 \times 21 \text{ bits} \times 30 \text{ meters} \times 39.37 \times 10^{-9}/\text{meter bit}}{1 - 90.66 \times t_s \times \lambda}$$

$$\bar{\tau} = \frac{744.093 \times 10^{-6} \text{ seconds}}{1 = 90.66 \times t_s \times \lambda}$$

t_s = number of bits x 30 meters x 39.37 x 10^{-9} seconds/meter bit

The values for λ = 10/second, 1/second, and 1/10 second are then obtained for messages ranging from 100 to 800 bits.

Configuration I computations: 64 processors

Given: The separation from IU 1 to IU 64 = 450 meters. Distance, d, is equal to 7.14 meters. A token has 21 bits. The speed of bit = 1 x 10^{-9} seconds/inch or 39.37 x 10^{-9} seconds/meter. The number of overhead bits/message = 42 bits.

Calculate:

$$\sum_{i=1}^{N} |i - (i+1)|, \text{ where } N + 1 \equiv 1; = 126$$

Equation (1) becomes

$$\bar{\tau} = \frac{126 \times 21 \text{ bits} \times 7.14 \text{ meters} \times 39.37 \times 10^{-9} \text{ meters/bit-sec}}{1 - t_s \times \lambda \times 1386.9}$$

$$\bar{\tau} = \frac{744.093 \times 10^{-6} \text{ seconds}}{1 - t_s \times \lambda \times 1386.9}$$

The values for λ = 10/second, 1/second, and 1/10 second are then obtained for messages ranging from 100 to 800 bits.

300 Modeling and Analysis of Local Area Networks

Configuration I computations: 64 processors

Given: The separation from IU 1 to IU 64 = 450 meters. Distance, d, is equal to 7.14 meters. A token has 21 bits. The speed of bit = 1×10^{-9} seconds/inch or 39.37×10^{-9} seconds/meter. The number of overhead bits/message = 42 bits.

Calculate:

$$\sum_{i=1}^{N} |i - (i+1)|, \text{ where } N + 1 \equiv 1; = 2048$$

Equation (1) becomes

$$\bar{\tau} = \frac{2048 \times 21 \text{ bits} \times 7.14 \text{ meters} \times 39.37 \times 10^{-9}}{1 - t_s \times \lambda \times 1386.9}$$

$$\bar{\tau} = \frac{12.08 \times 10^{-3}}{1 - t_s \times \lambda \times 1386.9}$$

The values for λ = 10/second, 1/second, and 1/10 second are then obtained for messages ranging from 100 to 800 bits.

Configuration I calculations: 16 processors

λ = 10/second

1. bits = 100 + 42 = 142; $\bar{\tau} = \dfrac{744.093 \times 10^{-6}}{.847949} = .8775 \times 10^{-3}$ seconds

2. bits = 200 + 42 = 242; $\bar{\tau} = \dfrac{744.093 \times 10^{-6}}{.74088} = 1 \times 10^{-3}$ seconds

3. bits = 400 + 42 = 442; $\bar{\tau} = \dfrac{744.093 \times 10^{-6}}{.526714} = 1.412 \times 10^{-3}$ seconds

4. bits = 600 + 42 = 642; $\bar{\tau} = \dfrac{744.093 \times 10^{-6}}{.3126} = 2.38 \times 10^{-3}$ seconds

5. bits = 800 + 42 = 842; $\bar{\tau} = \dfrac{744.093 \times 10^{-6}}{.099} = 7.5 \times 10^{-3}$ seconds

$\lambda = 1/\text{second}$

1. $142; \bar{\tau} = \dfrac{744.093 \times 10^{-3}}{.9848} = .755 \times 10^{-3}$ seconds
2. $242; \bar{\tau} = \dfrac{744.093 \times 10^{-3}}{.9526} = .7638 \times 10^{-3}$ seconds
3. $442; \bar{\tau} = \dfrac{744.093 \times 10^{-3}}{.9526} = .781 \times 10^{-3}$ seconds
4. $642; \bar{\tau} = \dfrac{744.093 \times 10^{-6}}{.9099} = .8176 \times 10^{-3}$ seconds
5. $842; \bar{\tau} = \dfrac{744.093 \times 10^{-6}}{.9099} = .8176 \times 10^{-3}$ seconds

$\lambda = 1/10 \text{ second}$

1. $142; \bar{\tau} = \dfrac{744.093 \times 10^{-6}}{.9848} = .755 \times 10^{-3}$ seconds
2. $242; \bar{\tau} = \dfrac{744.093 \times 10^{-6}}{.9974} = .7459 \times 10^{-3}$ seconds
3. $442; \bar{\tau} = \dfrac{744.093 \times 10^{-6}}{.9952} = .7475 \times 10^{-3}$ seconds
4. $642; \bar{\tau} = \dfrac{744.093 \times 10^{-6}}{.9931} = .749 \times 10^{-3}$ seconds
5. $842; \bar{\tau} = \dfrac{744.093 \times 10^{-6}}{.9909} = .7515 \times 10^{-3}$ seconds

Configuration II Calculations: 16 processors

$\lambda = 10/\text{second}$

1. $142; \bar{\tau} = \dfrac{3.1747 \times 10^{-3}}{.847} = 3.748 \times 10^{-3}$ seconds
2. $242; \bar{\tau} = \dfrac{3.1747 \times 10^{-3}}{.74088} = 4.285 \times 10^{-3}$ seconds
3. $442; \bar{\tau} = \dfrac{3.1747 \times 10^{-3}}{.5267} = 6.027 \times 10^{-3}$ seconds
4. $642; \bar{\tau} = \dfrac{3.1747 \times 10^{-3}}{.3126} = 10.15 \times 10^{-3}$ seconds
5. $842; \bar{\tau} = \dfrac{3.1747 \times 10^{-3}}{.099} = 32.06 \times 10^{-3}$ seconds

$\lambda = 1/\text{second}$

1. $142; \bar{\tau} = \dfrac{3.1747 \times 10^{-3}}{.9848} = 3.223 \times 10^{-3}$ seconds
2. $242; \bar{\tau} = \dfrac{3.1747 \times 10^{-3}}{.974} = 3.259 \times 10^{-3}$ seconds
3. $442; \bar{\tau} = \dfrac{3.1747 \times 10^{-3}}{.9526} = 3.3326 \times 10^{-3}$ seconds
4. $642; \bar{\tau} = \dfrac{3.1747 \times 10^{-3}}{.9312} = 3.409 \times 10^{-3}$ seconds
5. $842; \bar{\tau} = \dfrac{3.1747 \times 10^{-3}}{.9099} = 3.489 \times 10^{-3}$ seconds

$\lambda = 1/10$ second

1. $142; \bar{\tau} = \dfrac{3.1747 \times 10^{-3}}{.9984} = 3.179 \times 10^{-3}$ seconds
2. $242; \bar{\tau} = \dfrac{3.1747 \times 10^{-3}}{.9974} = 3.182 \times 10^{-3}$ seconds
3. $442; \bar{\tau} = \dfrac{3.1747 \times 10^{-3}}{.9952} = 3.19 \times 10^{-3}$ seconds
4. $642; \bar{\tau} = \dfrac{3.1747 \times 10^{-3}}{.9931} = 3.196 \times 10^{-3}$ seconds
5. $842; \bar{\tau} = \dfrac{3.1747 \times 10^{-3}}{.9909} = 3.203 \times 10^{-3}$ seconds

Configuration I calculations: 64 processors

$\lambda = 1/\text{second}$

1. bits = $100 + 42 = 142$; $\bar{\tau} = \dfrac{.744 \times 10^{-3}}{.944} = .788 \times 10^{-3}$ seconds
2. bits = $200 + 42 = 242$; $\bar{\tau} = \dfrac{.744 \times 10^{-3}}{.906} = .821 \times 10^{-3}$ seconds
3. bits = $400 + 42 = 442$; $\bar{\tau} = \dfrac{.744 \times 10^{-3}}{.828} = .898 \times 10^{-3}$ seconds
4. bits = $600 + 42 = 642$; $\bar{\tau} = \dfrac{.744 \times 10^{-3}}{.75} = .992 \times 10^{-3}$ seconds
5. bits = $800 + 42 = 842$; $\bar{\tau} = \dfrac{.744 \times 10^{-3}}{.672} = 1.107 \times 10^{-3}$ seconds

$\lambda = 1/10$ second

1. $142; \bar{\tau} = \dfrac{.744 \times 10^{-3}}{.9947} = .747 \times 10^{-3}$ seconds

2. $242; \bar{\tau} = \dfrac{.744 \times 10^{-3}}{.99} = .7515 \times 10^{-3}$ seconds

3. $442; \bar{\tau} = \dfrac{.744 \times 10^{-3}}{.982} = .7576 \times 10^{-3}$ seconds

4. $642; \bar{\tau} = \dfrac{.744 \times 10^{-3}}{.975} = .763 \times 10^{-3}$ seconds

5. $842; \bar{\tau} = \dfrac{.744 \times 10^{-3}}{.9672} = .769 \times 10^{-3}$ seconds

Configuration II calculations: 64 processors

$\lambda = 1/\text{second}$

1. $142; \bar{\tau} = \dfrac{12.08 \times 10^{-3}}{.944} = 12.79 \times 10^{-3}$ seconds

2. $242; \bar{\tau} = \dfrac{12.08 \times 10^{-3}}{.906} = 13.33 \times 10^{-3}$ seconds

3. $442; \bar{\tau} = \dfrac{12.08 \times 10^{-3}}{.828} = 14.58 \times 10^{-3}$ seconds

4. $642; \bar{\tau} = \dfrac{12.08 \times 10^{-3}}{.75} = 16.10 \times 10^{-3}$ seconds

5. $842; \bar{\tau} = \dfrac{12.08 \times 10^{-3}}{.672} = 17.97 \times 10^{-3}$ seconds

$\lambda = 1/10$ second

1. $142; \bar{\tau} = \dfrac{12.08 \times 10^{-3}}{.9947} = 12.14 \times 10^{-3}$ seconds

2. $242; \bar{\tau} = \dfrac{12.08 \times 10^{-3}}{.99} = 12.20 \times 10^{-3}$ seconds

3. $442; \bar{\tau} = \dfrac{12.08 \times 10^{-3}}{.982} = 12.3 \times 10^{-3}$ seconds

4. $642; \bar{\tau} = \dfrac{12.08 \times 10^{-3}}{.975} = 12.38 \times 10^{-3}$ seconds

5. $842; \bar{\tau} = \dfrac{12.08 \times 10^{-3}}{.967} = 12.49 \times 10^{-3}$ seconds

References

ACM Computing Surveys, Vol. 10, No. 3, September 1978.

Allen, A. O.: Probability, Statistics and Queuing Theory with Computer Science Applications, Academic Press, New York, 1978.

Computer, Vol. 13, No. 4, April 1980.

Fortier, P. J.: "Survey of LCN Architectures and Related Topics," NUSC Report, June 1980.

Fortier, P. J. and R. Leary: "Software Simulation Study of Local Computer Networks," NUSC Report, May 1980.

Gross, D. and C. M. Harris: Fundamentals of Queuing Theory, John Wiley, New York, 1974.

Kleinrock, L.: Queuing Systems, Vols. 1, 2, and 3, John Wiley, New York, 1975, 1976. + Kobayashi, H.: "Queuing Models for Computer Communication System Analysis," IEEE Trans. Comm., January 1977.

Leary, R.: "A Functional Description of Local Computer Networks," NUSC Report, July 1980.

Mendenhall, W., and R. Schaeffer: Mathematical Statistics with Applications, Duxbury Press, 1973.

Nagurney, A.: "Analytical Modeling Design Document of One Hughes IR&D Bus," Aquidneck Data Corp., August, 1980.

Nutt, G.: "A Case Study of Simulation as a Compute Design Tool," IEEE Computer, October 1978, pp. 31-36.

Page, E.: Queuing Theory in OR, Crane Russak, New York, 1972.

Trivedi, K., "Analytic Modeling of Computer Systems," Computer, October 1978, pp. 38-56.

Wagner, H.: Principles of Operations Research, Prentice-Hall, Englewood Cliffs, NJ, 1975.

Yuen, M. "Traffic Flow in a Distributed Loop Switching System," in Proceedings of Symposium on Computer Communications Networks and Teletraffic, Polytechnic Institute of Brooklyn, April 4-6, 1972.

Index

A

Aloha Net, 48
Analytical models, process and uses of, 6-8, 47
Arpanet, 12, 17, 48
Arrival rate, in queuing models, 129
Assembly line problem, simulations and, 109-110
Asynchronous timing, in simulation modeling, 89

B

Bank teller simulation example
    discrete event simulation, 90-91
    GASP IV, 95-98
    General Purpose Simulation System (GPSS), 100-102
    Simscript, 103, 104-105
    Slam II, 104-105, 107-109
Bayes' theorem, 54
Bernoulli trial, 69 Binomial distribution, 69-70
Birth and death processes, 120-123
Bottleneck analysis, 152, 178, 179
Bus network
    of local computer network, 221
    structure of, 23-25

C

Carrier sense multiple access, 32-33
Central moments, in probability analysis, 64-65
Central server model, queuing theory analysis, 163-170
Certain events, in probability theory, 52
Chebyshev's Theorem, 66
Chi-square distribution, 160
Chi-square test, in queuing estimations, 159
Closed networks, queuing models, 148-153
Communication lines, of local computer networks, 224
Computer network. *See* Local computer network
Conditional probability, 52-54
Confidence interval, in queuing estimations, 158-159
Contention-based protocols, 31-32
Continuous event simulation, 91-92
    triggering in, 92
    velocity of falling object example, 91-92
Continuous model, 8
Continuous random variable, 55, 58
Continuous system, 86
Controller, 30
Covariance, in probability analysis, 65-66

307

D

Database managers, 39-40
  levels of management, 39-40
Data transmission elements, 231
Decomposition, in queuing models, 128
DeMorgan's Law, in probability theory, 49-50
Densities, in probability theory, 58-61
Deterministic system, 86
Discrete event simulation, 89-91
  bank teller simulation example, 90-91
Discrete model, 8
Discrete random variable, 55
Discrete system, 86
Distributed database management systems, 39
  simulations and, 110
Distributed operating system, 38
Distributed Processing system, 198-204

E

Erlang density function, 82
Error detection, 231-233
  cyclic redundancy code check, 232
  feedback error control, 232
  forward error control, 231-232
  monitoring aspects, 232-233
  vertical redundancy check, 232
Error management, 40
Estimators, in queuing estimations, 156
Ethernet, 13, 48
Events, 233-235
  characteristics of, 233
  interface unit events, 234-235
  in probability theory, 49
  processor events, 234
  sequence of events in system flow, 235
Expectation, in probability theory, 61-68
Exponential distribution, 78-82

F

Flow balance equations, 148-150
Flow-equivalent servers, 128

G

GASP, 243, 255
GASP IV, 94-98
  bank teller simulation example, 95-98
General Purpose Simulation System (GPSS), 98-102
  bank teller simulation example, 100-102
Geometric distribution, 70-71
G/M/1 system, characteristics of, 146-147
Goodness of fit tests, 159

H

Honeywell experimental distributed processing system, 34
  analytical modeling of, 200-202
  graphic outputs, 202
  interface units, 198-200
Host processors, of local computer networks, 223
Hybrid network
  hybrid modeling, 94
  structure of, 28
Hypothesis test, in queuing estimations, 155-156

I

IEEE 802.5 token ring, 23
Interconnection structures, of local computer networks, 224-225
Interface units, 28-30
  of local computer networks, 223-224
  types of, 28-30
International Standard Organization, 229

J

Jackson's theorem, 155
Jointly distributed random variable, 55-56

Index   309

K

Kendall notation, 8
   use of, 131
Kolmogorov-Smirnov test, in queuing
   estimations, 160

L

Laplace transform, 146
Likelihood function, 158
Little's result, 170-171, 176, 177, 178
Local area networks (LANs)
   advantages of, 1
   bus network, 23-25
   evaluation for use of, 13-15, 43-47
   hybrid network, 28
   interface units, 28-30
   mesh network, 26
   modeling LANs, 44-47
   protocols, 30-37
   regular network, 26-28
   ring network, 22-23
   star network, 25
   systems management, 37-41
   uses of, 13, 14, 19-21
   compared to wide area networks,
      17-19
Local computer networks
   bus-structure, 221
   components
      communication lines, 224
      host processors, 223
      interconnection structures,
         224-225
      interface units, 223-224
      queues, 223
   definition of, 220
   events, 233-235
   network control, 220-221
   protocols, 226-231
      functions provided by, 227
      multilayer structure of, 226,
         229-231
      names/numbers of, 229

      problems related to, 228-229
      response types, 231
      transmission error detection, 231-233
   See also MALAN.

M

MALAN
   analysis module, 261-267
      establishing analysis criterion,
         262-265
      simulation and, 266
      statistical output, 266-267
   arbitrator module, 253-254
   arrival module, 245, 248-253
   design factors, 238
   functions needed, 237
   general model implementation, 239
   high-end data flow, 246-247
   implementation, 267-270
   implementation of model, 243-245
   interactive simulation interface,
      242-243
   IU evaluation parameters, 271-274
   mathematical presentation of
      statistics, 274-283
   message data types, 250
   next event simulation, 238-239
   periodic tabular data collection, 274,
      284
   requirements of distributed
      processing system, 241
   scheduling of events, 268
   simulator events, 268
   software implementation, 240-242
   structure of, 236-237
   topology module, 258-261
      ADDND, 260
      BUSTO, 261
      CDST, 258-260
      CNVRT, 260
      COLIDE, 260
      DIST, 260-261
      DVAL, 261
      RMVND, 260

use module, 245, 248, 255-261
   RECEPT, 257
   RESPRO, 257-258
   USEINIT, 257
   XMIT, 257
Manchester II scheme, 230
Markov process, 125-127
   Markov chain, 126, 145-146
Maximum likelihood estimation, in queuing estimations, 158
Mean, in probability analysis, 61
Mean value analysis, queuing theory analysis, 170-173
Mesh network, structure of, 26
Methods of moments, in queuing estimations, 157-158
M/G/1 system, characteristics of, 145-146
M/M/C system, characteristics of, 142-145
M/M/1 system
   characteristics of, 131-139
   queuing models, 131-147
M/M/1/K system, characteristics of, 139-142
Modeling tools
   analytical models, 6-8, 47
   operational analysis, 10-11, 48
   queue, 7-8
   simulation modeling, 8-9, 47
   test beds, 9-10, 48
Models, 2-6, 44-47
   benefits of, 44
   characteristics of good system, 3
   construction of, 3-6
   levels of analysis, 44-46
   measurement aspects, 45, 46-47
   methodology in, 4-6
   system in, 3, 86-87
   uses of, 3
Moments, in probability analysis, 64-65

N

Name management, 40
Network managers, 40
Network operating system, 37
Networks
   meanings of, 12
   reasons for use of, 12-13
Nodal activity, 235
Nonpersistent CSMA protocol, 33
Nth moment, 64
Null hypothesis, 156

O

Open networks, queuing models, 153-155
Open System Interconnect, 183
Operating systems, 37-38
   distributed operating system, 38
   network operating system, 37
Operational analysis
   operational theorem, 174
   operational variables in, 173-174
   performance quantities in, 175
   process and uses of, 10-11, 48
   queuing theory analysis, 173-179
Operationally connected network, 174

P

Performance evaluation parameters
   test beds, 187-190
   operational analysis, 188-190
Performance evaluation tools
   benchmarking, 195-196
      queuing models, 197-198
   examples of use
      calculations, 285-295
      Honeywell Experimental Distributed Processing system, 198-203
      token bus distributed system, 204-215
   linear projection method, 195-196
   rules of thumb, 195-196
   simulation, 196
   See also specific methods.
Permutations, in probability theory, 50
Persistent CSMA protocol, 33
Poisson distribution, 71-72
Poisson process, 118-120

Index

P-persistent CSMA protocol, 33
Probability/statistical analysis, 49-83
   axioms of probability theory, 51-52
   Bayes' theorem, 54
   certain events in, 52
   compliment of event in, 49
   conditional probability, 52-54
   DeMorgan's Law, 49-50
   densities, 58-61
   distribution function in, 56-58
   event in, 49
   expectation, 61-68
   normal curve values, table of, 77
   permutations in, 50
   probability distributions, 56-58
      examples of, 68-83
   random variables, 55-56, 58, 64-66
   sample spaces, construction of, 51
   starting point, 49
Product form, 165
Protocols, 30-37
   contention-based protocols, 31-32
   local computer networks, 226-231
      functions provided by, 227
      multilayer structure of, 226, 229-231
      names/numbers of, 229
      problems related to, 228-229
      response types, 231
   reservation-based protocols, 34-35
   sequential based protocols, 35-37

Q

Queues
   of local computer networks, 223
   process and uses of, 7-8
Queuing models, 8, 93
   arrival rate in, 129
   decomposition in, 128
   G/M/1 system, 146-147
   Kendall notation, use of, 131
   limitation of, 198
   M/G/1 system, 145-146
   M/M/1/K system, 139-142

M/M/1 system, 131-139
M/M/C/ system, 142-145
parameterization/distribution in, 155-161
   chi-square test, 159
   confidence interval, 158-159
   estimators, 156
   hypothesis test, 155-156
   Kolmogorov-Smirnov test, 160
   maximum likelihood estimation, 158
   methods of moments, 157-158
   sampling theorem, 157
probability/statistical analysis, 49-83
service discipline in, 131
service rate in, 130
steady state aspects, 130-131
stochastic processes, 117-125
Queuing networks, 128-129, 147-155
   closed networks, 148-153
   computational analysis
      central server model, 163-170
      mean value analysis, 170-173
      operational analysis, 173-179
   M/M/1 system, 131-147
   open networks, 153-155

R

Random variables
   Chebyshev's Theorem, 66
   continuous random variable, 55, 58
   covariance of, 65-66
   discrete random variable, 55
   jointly distributed random variable, 55-56
   nth moment, 64
   in probability theory, 55-56, 58, 64-66
Real messages, features of, 249-250
Regular network, structure of, 26-28
Request grant access, 33
Reservation-based protocols, 34-35
Ring network, structure of, 22-23
RS-232 protocol, 229

S

Sample spaces, construction of, in probability theory, 51
Sampling theorem, in queuing estimations, 157
Scan block, 198-199
Self-driven simulations, 88
Sequential based protocols, 35-37
Service rate, in queuing models, 130
Simscript, 103
    bank teller simulation example, 103, 104-105
Simulation languages, 9
    GASP IV, 94-98
    General Purpose Simulation System (GPSS), 98-102
    Simscript, 103
    Slam II, 103-109
Simulation modeling, 8-8, 85-87
    applications, 86, 109-110
    choosing level of simulation, 87-88
    combined modeling, 93-94
    continuous event simulation, 91-92
    discrete event simulation, 89-91
    hybrid modeling, 94
    process and uses of, 8-9, 47
    queuing modeling, 93
    reasons for use of, 85-86
    self-driven simulations, 88
    simulation program, 110-115
    steps in, 87-88
    time control in, 88-89
        asynchronous timing, 89
        synchronous timing, 88
        trace-driven simulations, 88
Slam II, 103-109, 112
    bank teller simulation example, 104-105, 107-109
Standard deviation, of random variable, 65
Star network, structure of, 25
Stationary type, stochastic processes, 118, 126
Statistical significance, 266

Steady state
    queuing models and, 130-131
    state transition diagram, 125
    state transition probability matrix, 126-127
    in stochastic processes, 123-125
Stochastic processes, 117-125
    birth and death processes, 120-123
    definition of, 117
    Markov process, 125-127
    Poisson process, 118-120
    state transition diagram, 125
    stationary type, 118, 126
    steady state in, 123-125
Stochastic system, 86
Synchronous timing, in simulation modeling, 88-89
Systems
    continuous system, 86
    deterministic system, 86
    discrete system, 86
    in modeling, 3, 86-87
    stochastic system, 86
Systems management, 37-41
    database managers, 39-40
    error management, 40
    network managers, 40
    operating systems, 37-38

T

Test beds, 181-194
    communication network in, 183-186
    message handling, 186
    performance evaluation parameters, 187-190
        operational analysis, 188-190
    performance tests, 190-194
        pipelining, 191
        utilization approaches, 190
    process and uses of, 9-10, 48
    prototype system, 181-182
    structure of, 183, 184
Token bus distributed system, 36

analytical modeling, case examples, 206-214
  bus use factor, 214-215
  computations, 297-303
  formulations/definitions in, 205-206
  graphic outputs, 215
Trace-driven simulations, 88
Triggering, continuous event simulation, 92

U

Utilization event, 235

V

Variance, in probability analysis, 65
Vectored protocol, 34-35
Velocity of falling object example, continuous event simulation, 91-92
Visit ratio, 152

W

Wide area networks, compared to local area networks, 17-19

X

X.21 standard, 229